THE SCIENCE OF JURASSIC PARK AND THE LOST WORLD

THE SCIENCE OF
JURASSIC PARK
AND THE LOST WORLD

OR,

HOW TO BUILD
A DINOSAUR

ROB DeSALLE
DAVID LINDLEY

BasicBooks
A Division of HarperCollins*Publishers*

OTHER BOOKS BY DAVID LINDLEY
The End of Physics
Where Does the Weirdness Go?

Published by BasicBooks,
A Division of HarperCollins Publishers Inc.

Designed by Laura Lindgren

Library of Congress Cataloging-in-Publication Data
  The science of jurassic park, or, How to build a dinosaur
/ Rob DeSalle and David Lindley.
      p.   cm.
   Includes index.
   ISBN 0-465-07379-4
      1. Genetic engineering. 2. Jurassic Park (Motion picture).
3. Dinosaurs. I. Lindley, David, 1956–   . II. Title.
QH442.D47   1997
660'.65-dc21   97-660
                             CIP

97 98 99 00 01 ❖/RRD 10 9 8 7 6 5 4 3 2 1

CONTENTS

ACKNOWLEDGMENTS

We would like to thank Susan Rabiner, our editor at Basic Books, for initiating and guiding this project, Pamela Hines for suggestions and advice throughout, and Sara Lippincott for turning an unruly manuscript into a trim narrative. We also thank David Grimaldi, Dennis Finnin, Mick Ellison, and Edward Heck for the illustrations, and John Fleming for the clarifications.

David Lindley also wishes to thank Annie Fortunato, Karen Hopkin, Rich Monastersky, Liz Pennisi, and John Travis for a variety of ideas and explanations.

Rob DeSalle would like to thank colleagues David Grimaldi for the permission to use his extraordinary photographs and Ward Wheeler and Mark Norell for advice and information. This book would not have been possible for me without the support, advice, and companionship of Annie Williams.

PROLOGUE

ake a seat. The tape's in the VCR, the remote control's in your hand, the popcorn's at your side. Get comfortable. Relax. Hit the play button. You're watching Steven Spielberg's movie of Michael Crichton's best-selling novel Jurassic Park. *Across the bottom of the screen, a caption tells you where you've been transported to:* "Mano de Dios Amber Mine, Dominican Republic." *A Caribbean island. As you're drawn into the story, you imagine yourself part of the action, hanging around with the actors and the crew as the movie is being made.*

You're on a rickety raft, being pulled across a shallow river to a flat, muddy landing area. It's hot, oppressive. You cautiously step off the raft and find yourself in the middle of what appears to be a large, shallow excavation dug out from the surrounding dense forest. In the small cliffs that ring this quarry, you see timber-framed openings. Through them, miners are coming and going with baskets. You move forward to investigate.

Now you're in one of the caves in the cliffside. Around you, local workers are pounding away at the cave walls, looking for pieces of amber buried in the soft, flaky rocks.

Uh, Mr. Spielberg? Sorry to interrupt, but, uh, you're not likely to find any amber here, no matter how far into the riverbank you dig. Dominican amber mines are mostly up in the hills, away from the rivers. No jungle. More like a sort of open scrub—you know, little trees and shrubs, grass, that kind of thing. And most of the amber mining isn't done in caves. . . . No, I guess you've got a point—it wouldn't be nearly as scenic. But we want to do things right, don't we? We just have to get the actors and the crew together and move everything upland a few miles. OK?

Now you're away from the river, on an unkempt hillside. Miners are hacking the ground with shovels and machetes, leaving open, ugly scars. The rocks beneath the thin soil break apart under the blows, and among the fragments the practiced eyes of the miners pick out darker flecks. These are pieces of petrified wood, looking almost like coal, and in them amber is often found. Amber starts out as a kind of gum that oozes from trees when the bark is wounded; over time, it hardens into a yellowish, resinous glob, looking almost like a piece of plastic. Dominican amber comes mostly from timber trees of the genus *Hymenaea*. The wood of these trees is soaked through with resin; it helps protect the tree from colonies of beetles that like to tunnel into the wood. In Central America today, the branches of *Hymenaea* are heavily coated with resin, looking like honey-dippers just pulled from the pot and trapping all kinds of incautious flies and mosquitoes. The hardened resin is astonishingly tough. Once it's tucked safely underground, away from air and water, it can lie undisturbed and uncorroded for thousands or millions of years, even 100 million years or more. . . .

Occasionally, the workers will come across a trail of fossilized wood fragments, remnants of a huge tree trunk that fell millions of years ago into a muddy swamp. Over the ages, the thick muck of the swamp becomes compressed and compacted, and turns eventually into rock. Within that rock are preserved pieces of the tree trunk, along with globs of tree resin that have solidified into amber.

Following a rich vein of amber-containing fragments, the miners tunnel into the hillside, creating a cave—

That's good! A cave!

—that may yield a treasure trove of amber pieces. The miners know this area well. They make a living by digging amber and selling it to local traders, who clean the rough pieces and

Reconstruction of a *Hymenaea* forest from the Dominican Republic about 25 million years old.

(Drawing by Dr. David Grimaldi)

sell the brightest and clearest of them to collectors around the world. Translucent amber jewels and ornaments have been in demand since the time of the ancient Egyptians. A shiny amber ornament or a necklace of beads can sell for hundreds of dollars.

But you're here for a different reason. You're looking for something rarer and more valuable than mere decoration. As the miners pick over what they find, one comes toward you with an unusually large piece of amber—large enough to nestle comfortably in the palm of your hand. Cleaning the surface, the miner holds his find up to the light. Embedded in the bright yellow nugget, which still seems to glow with the sunlight that bathed it on the day it was formed, is a huge, perfectly preserved insect, suspended unchanged and looking as if it might spring to life and fly off if you were to gently crack open its glowing yellow sarcophagus.

Oh, Mr. Spielberg, I'm really sorry to interrupt again, but I just remembered something important. This Dominican amber certainly does have insects in it once in a while, but they're not going to be any good for what we want. They're about 25 million years old, maybe 35 million. That's the age of the rock the amber is in, so that's how long ago these particular trees died. . . . You're right, that is pretty old, but it's just not old enough. Dinosaurs went extinct some 65 million years ago. So this little insect may have some sort of animal blood inside it from the last time it fed, but it can't have any dinosaur blood, because dinosaurs were long gone by the time this particular insect got trapped in resin. . . . Uh, no, I'm afraid the amber here is all about the same age. . . . Mexico? Sorry, that's no good either. All the amber in Central America is pretty much the same age. We're going to have to pack up everything and move the whole story to a place where you might find amber that's 65 million or more years old.

Wait! I remember reading recently that a scientist at the American Museum of Natural History in New York made this wonderful discovery—exactly the kind of thing we need! A mosquito trapped in amber that was at least 85 million years old. Plenty of dinosaurs around then. No one knows if this mosquito actually has any di-

nosaur blood in it—but it could, certainly! . . . Yes, it's perfect, isn't it? Just what we wanted. . . . Where? Ah, well, that's the bad news, Mr. Spielberg. They found it in New Jersey. . . . I know—if you'd set Raiders of the Lost Ark *in Hoboken, it wouldn't have lasted a week at the theaters. But we really do want to get this right, don't we? Off to New Jersey it is, then!*

It's a sweltering East Coast summer. The noise and the fumes of traffic and construction machinery are all around you. You're in a field, up to your ankles in thick, gluey clay. A year from now, this particular field will be a shopping mall, but before the site is covered in concrete, you and your team have the chance to dig up the soil looking for amber. It's hard, dirty work. You turn over a spadeful of the heavy clay and rummage through it by hand. There are rocks and stones, bits of fossilized wood, and every now and then a piece of amber. It's hard to pick out pieces of amber in all this muck; they don't look shiny and bright until they're polished. For the moment, they are dull, coarse, brownish lumps—nothing out of the ordinary. But they feel light in your hand—almost light enough to float.

In an afternoon of backbreaking work, you come up with a couple of dozen small pieces. Many are tiny, like little bits of shattered glass that have been worn down and rounded over the years, but you come across a few marble-size pieces too. Right now, you can't tell whether there's anything interesting here. You have to get the amber back to your lab, where you can clean it and polish it to restore its bright, orange-yellow translucence.

Once there, you examine the pieces through a microscope. You see countless midges and tiny worms, not to mention fragments of plants and flowers, but bigger insects are rare. In the 80 pounds of amber you've collected over the summer, you will find only one mosquito.

Still, now you have what you were looking for. A biting insect, 85 million years old, that was living and breathing and—most important—feeding in a world dominated by dinosaurs. It amazes you to think that this very mosquito once flew among dinosaurs. And it may, you hope, even have lunched on dinosaur blood. If you are really lucky, it may have fed not too long before it was trapped on the sticky bark of some ancient tree and swallowed up in the glob of resin that hardened into the piece of amber you are holding. You can see that the amber has preserved the insect with unbelievable perfection. Under the microscope, tiny details of its wing structure are easily seen. You can make out little hairs on its back and legs, the most minute features of its head.

Now all you have to do is open this nugget of amber and see what you've got. . . .

INTRODUCTION

TERRIBLE LIZARDS

Say "Jurassic Park" and people instantly know what you're talking about. Since 1990, when it was published, some 10 million people have bought the book. The movie has grossed over $350 million since its release in 1993. And these readers and viewers can tell you without hesitation what the story is about: the inventive scientists of John Hammond's laboratories on Isla Nublar bring dinosaurs back to life, the dinosaurs turn out to be smarter and scarier and harder to control than anyone suspected, and mayhem ensues. In the end, the good guys survive, leaving behind them the violently trashed remains of what was supposed to be a magnificent international tourist attraction off the coast of Costa Rica.

You could call it just another version of the old Frankenstein tale: well-intentioned scientist creates living creature, creature becomes monster, monster goes on the rampage. That scenario, no matter how many the variations on its theme, still has the power to fascinate and terrify. But the extraordinary success of *Jurassic Park* is due in part to Michael Crichton's addition of a couple of ingenious twists to the story.

First, there's the idea that bringing dinosaurs back to life may someday—thanks to the dizzying progress of twentieth-century genetics—become science fact rather than science fiction. In 1912, Arthur Conan Doyle (known as the creator of Sherlock Holmes) wrote a fantastic piece of fiction called "The Lost World." In this tale, the intrepid Professor Challenger travels to darkest South America and discovers a lost plateau, high in the Andes, isolated from the rest of the world, and inhabited by dinosaurs. In Conan Doyle's day, you could still think of South America as a mysterious, unknown place, where ancient beasts and wonders might well await the determined explorer. Though Crichton paid his respects by appropriating his predecessor's title for the sequel to *Jurassic Park*, he apparently realized—as did a number of other science fiction pioneers, like H. G. Wells and Jules Verne—that beasts and wonders are more likely to dwell in the unknown regions of science than anywhere on the teeming and thoroughly explored earth.

Resurrection of creatures from our planet's prehistoric past no longer seems an impossibility. Modern scientists have already recovered ancient human genetic material. They have found bits of DNA—deoxyribonucleic acid, the lengthy helical molecule that carries our genetic makeup—from Egyptian mummies, and also from an unfortunate fellow who died in the Alps thousands of years ago and became trapped in ice. Bits of DNA from an extinct cousin of the zebra have been recovered from scraps of dried skin in a Berlin museum. A number of scientists have succeeded in extracting fragments of DNA from insects and plants that died around 30 million years ago. Even more exciting, one researcher—George Poinar, of the University of California at Berkeley—was able to tease out a tiny piece of the genetic material from a beetle he found in amber from Lebanon—amber thought to be some 125 million years old and, therefore, from a time when the dinosaurs still thrived. These are real scientific accomplishments, breakthroughs of the last decade or so. Is it therefore so unreasonable to imagine that some years from now dinosaur DNA might be coaxed out of an ancient remnant?

And second, of course, there are the dinosaurs themselves. Space aliens, malevolent robots, zombies, coiled behemoths from the depths of the sea—all these have figured time and time again in books and movies. But they're inventions, fictions. Dinosaurs are—or were—real. Tens of millions of years ago, long before humans (or, indeed, any large mammals at all) appeared on the scene, dinosaurs ruled the planet. The biggest of them were terrifyingly large, and some were undoubtedly ferocious predators. Just to think of those creatures majestically treading the same ground we walk upon today inspires awe, and perhaps a sense of our own fragility. Numerous species of dinosaur inhabited the earth through three long periods: the Triassic, from 250 to 210 million years ago; the Jurassic, from 210 to 140 million years ago; and the Cretaceous, from 140 to 65 million years ago. Of the dinosaurs inhabiting John Hammond's mega-theme park, more species belonged to the Cretaceous period than to the other two periods, so it's actually a bit of a stretch to call the place "Jurassic Park." It ought to be named Cretaceous Park, if anything—indeed, some of the more spectacular animals, such as *T. rex* and the fearsome velociraptors, did not put in an appearance until late in that period, missing the Jurassic altogether!

What is it about dinosaurs that beguiles us so? Their sheer immensity, the image of something both huge and alive, must be a large part of the explanation. But dinosaurs also force us to think of immensities in time. The reign of dinosaurs on earth lasted 185 million years, far longer than the paltry 3 or 4 million years that humans and their hominid ancestors have been scratching out an existence. The dinosaurs must be counted successful creatures, and yet they died out. The last dinosaurs disappeared about 65 million years ago, and in their wake other creatures came along—smaller creatures, more ingenious and more adaptable, perhaps, but nowhere near as grand.

From the first discovery of dinosaur fossils in the early 1800s, there was an irresistible tendency to think of dinosaurs as dull, lumbering giants, dominating the earth by brute force alone. The scientists of the time, along with everyone else, regarded humanity as the peak of creation, so these colossal creatures, now extinct, must necessarily have been slow-witted, and successful only because of dumb luck. All kinds of imaginative hypotheses were dreamed up to explain their extinction, most of them unflattering to the dinosaurs. The climate changed and they were too stupid to move somewhere warmer; or the vegetation changed and they all became constipated; or—a particularly anthropocentric twist—the first mammals came along and these tiny, scuttling, clever creatures stole the dinosaurs' food from under their noses. If dinosaurial bulk made the dinosaurs successful for a while (most of the theories went), it also made them sluggish and complacent, fatally slow to respond to change. In the Victorian era, people could feel comfortably superior to these extinct brutes. Earlier forms of life were by definition failures—rough drafts of things to come—and humanity was God's finest handiwork.

But in the last few decades our view of dinosaurs has undergone a subtle change. For one thing, it used to be thought that dinosaurs were cold-blooded, and this idea carried with it all the psychological implications of cold-bloodedness. When the pioneering British paleontologist Richard Owen was examining the first dinosaur fossils, he observed that although their skeletons had something in common with those of birds (especially in the formation of the hips), they also resembled crocodiles. In 1842, he coined the term "dinosaur," from the Greek for "terrible lizard," to describe this new class of creature, and the name stuck—with a vengeance. Dinosaurs were thought of as reptiles—like modern lizards, snakes, and crocodiles—and for many decades the effort to understand dinosaurs and the world they lived in was predicated on their presumably cold-blooded, reptilian nature.

Birds are feathery and warm-blooded, dinosaurs were scaly and cold-blooded—or so the old orthodoxy maintained. But in recent years the unquestioned assumption that the dinosaurs were cold-blooded has come under scrutiny. As the paleontologist Robert Bakker argues in his lively 1986 book *The Dinosaur Heresies,* it's not that anyone ever produced any evidence of dinosaurs' cold-bloodedness, but that they were classified from the beginning as reptiles and cold-bloodedness is characteristic of reptilehood.

Why is the distinction between cold-bloodedness and warm-bloodedness so important? Mostly because it determines your lifestyle. Warm-blooded animals, such as ourselves, maintain a constant body temperature by means of our metabolism; we keep ourselves warm from the inside out. Cold-blooded animals rely on external heat. When it's hot outside, lizards bask in the sun, soaking up heat; when it turns cool at night, or during a stretch of cold weather, they slow down. Their metabolism goes up and down with the weather. If you're warm-blooded, you have to eat more than a cold-blooded creature of the same size, because you have to keep your internal fires stoked. In evolutionary terms, this is a disadvantage: finding food takes effort, so you have to work harder to stay alive. The evolutionary payoff, however, is that warm-bloodedness *allows* you to work harder. You can run after prey, or away from predators, at all hours of the day and night and in all seasons. A cold-blooded creature, on the other hand, has a limited capacity for chasing its food, and if something big and fierce comes across it during a cold spell, it lacks the energy to get away. One argument against dinosaurs as warm-blooded creatures is that a big dinosaur—the 20-ton apatosaurus, for example—would have been tremendously vulnerable. You could imagine an apatosaur, forced into semidormancy during a spell of cold weather, being nibbled to

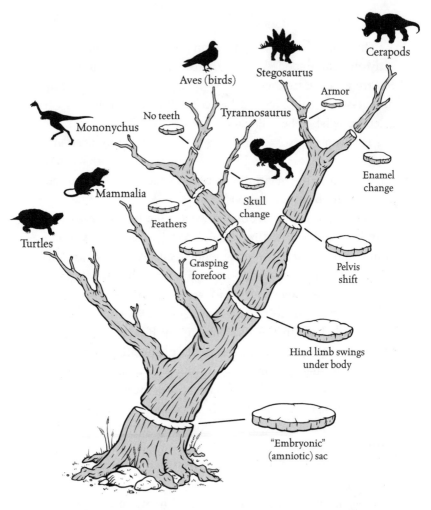

An evolutionary tree showing the close relatedness of dinosaurs and birds. Cross sections are cut out of the tree and represent the major anatomical innovations that the common ancestor of all animals above the cross section developed during evolution.

(*Drawing by Edward Heck*)

death by scurrying, cat-size mammals. On account of its size, the apatosaur would have had no place to hide and no energy to fend off its pesky attackers.

This sort of argument—that dinosaurs must have been warm-blooded to function successfully, as they evidently did—was advanced in the 1960s by the Yale paleontologist John Ostrom. He was particularly taken with the smaller predatory dinosaurs, such as velociraptors, which, he argued, must have been fast, fierce attackers, capable of swift locomotion and agile maneuvering. To be successful, able to give chase whenever the opportunity presented itself, such dinosaurs must have been warm-blooded, Ostrom argued. Bakker also took up the cause of warm-bloodedness, seeing in the fossil evidence other behaviorial and physiological traits: the dinosaurial diet, rate of growth, ratio of carnivores to herbivores—all suggested to him that the dinosaur lifestyle was closer to that of modern mammals than of reptiles. (Steven Spielberg appears to agree: in the movie of *Jurassic Park*, the raptors invade a refrigerated room in pursuit of Hammond's grandchildren and exhale foggy breaths, as you or I, being warm-blooded creatures, would do on a cold day.)

But many dinosaur experts, probably the majority of them, remain less than convinced by the arguments that Ostrom and Bakker have put forward. Modern crocodiles, these critics point out, are most certainly cold-blooded but nevertheless produce bursts of energy sufficient to catch their prey. How can the debate be settled? One hope is that fossil bones will ultimately settle it. Cold-blooded animals grow intermittently, with the seasons, while warm-blooded animals grow at their own fairly constant rate. The bones of modern reptiles thus tend to show growth rings something like the seasonal growth rings of trees, while those of mammals generally do not. And the bones of mammals are pervaded by a dense network of blood vessels to supply nutrients, while reptile bones tend to have simpler structures. But detailed examination of fossilized dinosaur bones has provided ammunition for both sides of the debate: some appear to show the complex internal network of capillaries typical of warm-blooded creatures, but others show growth rings. And then there's the fact that almost all modern warm-blooded animals have cartilaginous structures in their nasal passages, the purpose of which is to reduce loss of heat and moisture as the animal breathes. So far, no dinosaur skull has shown evidence of any such structure.

Meanwhile, careful excavation of fossil sites around the world has yielded the remains of dinosaur nests and other indications that these beasts—cold-blooded or not—formed family and social groups. In the early part of this century, the paleontologist Roy Chapman Andrews, of the American Museum of Natural History in New York, led an expedition to the red sandstone formations of Mongolia and dug up fossil evidence of such nests. More recent expeditions to Mongolia, led by Mark Norell, Andrews's equally adventurous successor at the museum, uncovered an intact fossilized nesting site with the bones of a dinosaur still perched on its eggs—perhaps trying to protect the nest from a fierce sandstorm that killed it and its unhatched offspring. Ironically, the species of dinosaur on this nest had earlier been named oviraptor—Latin for "egg stealer"—because the first specimen was found close to a clutch of eggs, upon which it was assumed to be preying. It nows appears that the oviraptor was instead an exemplary parent. And in the badlands of Montana, the paleontologist Jack Horner has found vast areas that were evidently the breeding sites and adolescent playgrounds of great herds of a hitherto unknown species of plant-eating dinosaurs. The disposition of these fields of fossils indicated that when the babies were hatched they were unable to take care of themselves, and that even after the young had left the nest they were looked after by their parents. Horner named his new dinosaur species *Maiasaurus*—"good-mother lizard."

These were, it now seems, creatures with complex social behaviors, not dumb killers or hopelessly stupid leaf munchers. And although dinosaurs were once thought to be a dead end of evolution, many scientists believe that birds, which must surely be counted among the most successful animals on earth, are the distant descendants of one particular branch of the dinosaur family. Dinosaurs themselves may be gone, but their legacy is not entirely lost. One of the most thrilling of fossil discoveries came in Bavaria in 1861, with the unearthing of *Archaeopteryx* ("ancient bird"), a creature that had the teeth, claws, and general skeletal appearance of a small dinosaur but also clearly possessed feathered wings. (This fossil, along with a couple of others later unearthed in the same area, was found in unusually fine sedimentary rocks, which had preserved the feathers in impressive and convincing detail.) Over millions of years, random mutation would have led to the appearance of longer and longer feathers and, ultimately, to flight.

There is still some debate over whether *Archaeopteryx* could fly or merely clambered into trees and glided clumsily to the ground. (Interestingly, a strange and primitive South American bird called the hoatzin retains claws at its wingtips and climbs in trees.) No matter: *Archaeopteryx* seems a very probable precursor of modern birds, and a few other species of fossilized dino-birds have since been found. *Archaeopteryx* dates back to 150 million years ago, which means that birds developed as an offshoot of dinosaurs long before the great extinction of 65 million years ago. Throughout the Cretaceous period, birds of a reasonably modern appearance might have been flitting around among the dinosaurs.

This evolutionary hypothesis has its detractors: not every paleontologist agrees that birds are the direct descendants of dinosaurs—or that *Archaeopteryx* was the ancestor of modern birds. Very recently, fossils of what appear to have been primitive birds—animals much older than *Archaeopteryx*—were found in China. If they are indeed as old and as birdlike as their discoverers suggest, then they would have branched away from dinosaurs at a much earlier stage, making them dinosaur cousins, like snakes and crocodiles, rather than dinosaur descendants. In that case, *Archaeopteryx* would have been merely an evolutionary side street that became a dead end.

Finally, any comfort we might have taken from thinking that the dinosaurs died because of their supposed stupidity or clumsiness has been upended in recent years.

It's impossible to tell by studying dinosaur fossils alone whether the dinosaurs disappeared over millions of years or literally overnight. Big fossils are comparatively rare, so it's not a question of seeing hundreds of dinosaur bones in one layer of rock and none at all in an overlying (and therefore younger) layer. At one site, paleontologists might find a tyrannosaur thighbone about 68 million years old and somewhere else a couple of vertebrae about 66 million years old. They find nothing at all younger than 65 million years old. But does that mean that tyrannosaurs all died suddenly 65 million years ago or that they gradually dwindled to nothing over an unknown period of time? For such reasons, there was a long debate over whether dinosaurs died out suddenly or slowly. Many paleontologists were more comfortable with the idea of gradual decline, but now it is almost universally accepted that the dinosaurs were killed off by a catastrophe they could do nothing about.

The picture began to change in the late 1970s, when the Nobel Prize–winning physicist Luis Alvarez and his son Walter, a geologist, began studying some curious rock formations in Italy. They analyzed a distinct dark layer of rock, a few millimeters thick, at the geological boundary between the Cretaceous and the Tertiary periods—about 65 million years ago—and found it to contain a high proportion of iridium, a metal uncommon on earth but found abundantly in meteorites that reach the earth's surface from outer space.

To cut a long and controversial story short, Alvarez and Alvarez proposed that a huge rocky object, possibly a stray asteroid, struck the earth violently and disastrously one day about 65 million years ago. The impact of this body, which they estimated must have been about 10 miles across—bigger than the island of Manhattan—would have created a huge crater and sent clouds of dust flying into the atmosphere, with effects both devastating and global. The dust would have been picked up by stratospheric air currents and spread around the world, blotting out the sunlight; the planet would have suddenly become cooler and darker; plants would have died. Dinosaurs, not to mention a lot of other animals, would have found themselves shivering and hungry in a cold and gloomy world. It would take decades, perhaps even centuries, for the dust to settle and the climate to recover. And in that time, the dinosaurs died.

This was a contentious hypothesis, to say the least. Although the rather modest amount of dust put into the atmosphere by the volcanic eruption of Mt. Pinatubo in the Philippines in 1991 spread across the globe, causing some unusually pink sunsets and a temporary cooling of the atmosphere by a fraction of a degree or so, this and other such documented phenomena are nothing compared to what the Alvarezes proposed had happened at the end of the Cretaceous. Nevertheless, the idea began to gain support. The iridium-rich layer was found all over the earth, and always in geological layers corresponding to an age of 65 million years. Scientists came up with other ways to determine the suddenness of the extinction. Tracking the number of pollen grains in rock just above and just below the telltale iridium layer, paleobotanists found a marked and abrupt change: there was a lot of pollen below (that is, before the supposed impact) and much less pollen above (that is, after the impact). It wasn't just dinosaurs that died out—a large fraction of the earth's plant life disappeared as well.

The Cretaceous extinction was truly a mass extinction. Dinosaurs disappeared along with many kinds of plants, marine creatures, and smaller land animals—including a number of mammals. There are still a few scientists who hold out against the giant impact hypothesis, but on the whole it's gone from being a far-out idea, widely mocked when it was first proposed, to the standard explanation. A final piece of evidence for the impact theory was the discovery of the place where the impact seems to have occurred. Although the crater would have been enormous, 65 million years of erosion and deposition would have largely worn it down and filled it in. Meteor Crater, in Arizona, remains quite distinctive in appearance, but it's estimated to be no more than a few tens of thousands of years old. (Meteor Crater is something less than a mile across, and the meteor that made it was probably only about 20 yards in diameter—a good deal smaller than the object proposed by the Alvarezes.) Still, there would have been evidence in the subsurface of a catastrophe of that magnitude. In 1991, geologists scrutinizing subsurface rock formations in the Yucatán Peninsula of Mexico announced that they had found violently distorted layers of rock. The pattern of these distortions formed a circular structure more than 100 miles across, and everything about this ancient crater—notably, its size and age, and scattered evidence of unimaginable heat in the form of glassy globules—suggests that here is the site of the Cretaceous catastrophe. There really was a huge impact 65 million years ago, and it happened in what is now Mexico.

That doesn't mean the debate is over. How exactly did the

impact fulfill its disastrous role? Dust blotting out the sun is certainly a likely possibility, and the impact itself might well have been felt around the world, perhaps setting off secondary earthquakes and volcanoes to add to the damage. Or, perhaps, the impact sent not just dust but also noxious gases up into the atmosphere, some of which could have destroyed the ozone layer. Take away the protective ozone and the earth's surface would have been bathed in harmful ultraviolet: dinosaurs could have died because of damaged eyesight or skin cancers. The violent climatic disruption might also have spawned super-hurricanes, thousands of miles across, which scoured the planet's surface.

How did other animals—most particularly, the tiny mammals that eventually gave rise to much of the modern world's animal life, including ourselves—manage to survive? The most likely answer seems to be that they were more adept at eking out a living in harsh circumstances than were the stately dinosaurs, which had lived so long in relative comfort. It has also been suggested that the earliest mammals were nocturnal, because it would have been easier for them to survive among the dinosaurs by coming out only at night. If so, their nocturnal habits would have been an advantage in the cold, dark, world-wide winter after the impact. In the postimpact world, little scavenging rodents—the ancestors of modern mice and rats—would have done better than the maiasaurs, for example, who were accustomed to grazing on lush, abundant vegetation. And if the maiasaurs disappeared, what would *Tyrannosaurus Rex* have had for dinner? Hard to imagine the king of the dinosaurs surviving on mice and mushrooms, and picking for bugs under fallen trees.

And of course there's always the question of what would have happened if this stray asteroid had swung just 1,000 miles to one side and missed the earth altogether. At one point in *Jurassic Park*, the "chaotician" Ian Malcolm mutters that nature gave dinosaurs their chance and they ultimately failed, and that it wasn't John Hammond's business to give them another go-round. But you could argue that mass extinction by asteroid impact has little to do with the normal course of nature—nature on earth, at any rate—and that the disappearance of the dinosaurs is the unfortunate consequence of an arbitrary cosmic accident. Whether you put the dinosaurs' demise down to bad luck or a lack of adaptability, we humans should probably be thankful that it happened. As the old saying goes, it's an ill wind that blows nobody good.

But will scientists really be able to re-create dinosaurs one day? If you can find bits of DNA from 125-million-year-old insects, why not from 65-million-year-old dinosaurs? If you can find the DNA of a dinosaur, have you not found its genetic blueprint, the biochemical recipe for making the animal itself? The idea of building a dinosaur may be altogether speculative, to be sure, but it has the ring of speculation built on genuine scientific achievements and knowledge.

On the other hand, you don't have to be a scientist to guess that there might be a bit more to the task than that. If the cold war were to be fought again, this time (for some bizarre reason) with the Russian and American governments deciding that world domination would go to the first nation to resurrect an authentic dinosaur, there's no way either side could put together a team of expert scientists with a few billion dollars and expect the job to be done in a matter of years. The Manhattan Project of the Second World War, which resulted in the first atomic bomb, was a huge enterprise, but it was based on established scientific fact: even though enormous difficulties were involved, it was pretty clear from an early stage that such a bomb could be made in principle. A Manhattan Project to build a dinosaur would be a different matter altogether.

There's an awful lot about DNA that scientists still don't understand. The process by which a chicken's egg, say, turns chicken DNA into a whole new chicken is only dimly understood. A lot of fundamental scientific problems would have to be solved before the idea of turning dinosaur DNA into an actual dinosaur could even be usefully tackled, let alone accomplished.

What's more, the task of building a dinosaur is not one big problem but a host of problems small and large. Some apparently insignificant detail, some tiny piece of missing information, could undo the whole enterprise. Or you might find yourself beset with a multitude of niggling difficulties—snags you might be able to deal with individually but which together overwhelm your abilities and resources. Even if the science of *Jurassic Park*, at least in broad outline, is based on fact and on plausible guesswork, that doesn't guarantee that you'll be able to solve all the problems that will inevitably arise.

But then again, although any number of these problems might be insoluble now, it's not clear that any of them are insoluble in principle. Genetics and molecular biology are, comparatively speaking, in their infancy. There's a huge amount that is, as scientists are fond of saying, "not well understood" yet. This book, despite its subtitle, does not guarantee that when you get to the last page you'll know how to build a dinosaur. It doesn't even claim that you'll be able to decide whether building a dinosaur will ever be possible or not. But you will, we hope, have a better idea of what sort of problems stand in your way and how much work you'll have to do to solve them.

CHAPTER ONE

DIGGING FOR AMBER

So you want to make a dinosaur?

First things first, then. As far as anyone knows, the stuff dinosaurs were made from was no different from what we and all other living creatures are made from—proteins, fats, carbohydrates and sugars, lots of water, a dash of salt, iron for the blood, calcium for the bones, various other minerals here and there for special purposes. It's how all those things are put together that counts.

What you really need is the set of instructions for making a dinosaur. A blueprint, if you like. To scientists, that blueprint is known as the animal's genetic code, or genome—its DNA. You, a human being, have human DNA—your complete genome—in just about every cell of your body: skin cells, muscle cells, liver cells, nerve cells. You could get a sample of your own DNA from a mere scraping of skin. But how do you acquire the DNA of an animal that became extinct 65 million years ago?

In Michael Crichton's story, Lewis Dodgson—the unscrupulous product-development chief for the Biosyn Corporation of Cupertino, California—hears rumors that scientists at the rival Ingen Corporation have acquired some dinosaur DNA, and he speculates that they have extracted it from fossils. It used to be thought, he muses, that "fossilization eliminated all DNA. Now that was recognized as untrue."

Is Dodgson onto something? Almost everything we know about dinosaurs comes from the study of dinosaur fossils. Could Ingen's biogeneticists have got DNA from fossils too?

On an upper floor of the American Museum of Natural History—a precinct not normally visited by the public—there's a big, poorly lit, dusty room lined with plain metal shelves on which sit all kinds of unsorted and unclassified fossils, large and small, covered by plastic sheets. The fully assembled dinosaur skeletons you see on public display downstairs were not found that way in nature. Intact remains are rare; many of these displays have been put together from bits and pieces discovered in many different places. The museums of the world contain all sorts of broken remnants, useful to researchers but not so interesting to the average museumgoer. Why not just gather together a few truckloads of dinosaur fossil fragments, grind them all up into powder, and see whether or not some DNA was preserved somewhere in them? You need only a single complete copy of the dinosaur's DNA, and a pound of dinosaur remains would have contained something like a trillion individual cells, each containing one copy of the animal's DNA. Is it too much to ask that in a ton of otherwise unexciting fossil fragments just one copy of the DNA, out of all those trillions, could have survived unchanged?

As a matter of fact, the answer is, Yes, it probably is too much to ask.

When an animal dies, its corpse usually doesn't stick around for more than a few months. Nature works quickly, with effects all too obvious to anyone passing by. Many creatures (for example, ourselves) are repelled by the odor of decomposing flesh—an evolutionary development that protects us from eating bad meat. Scavenging animals, on the other hand, have evolved to recognize those same putrid smells as the indication that something yummy is nearby, and they flock to the feast. Insects and bacteria take care of most of what's left over. Chemical reactions cause complicated organic molecules to break down into smaller, simpler, and often

This dusty assemblage of dinosaur fossils resides two floors above the fourth floor dinosaur halls at the American Museum of Natural History. (*Photo by Dennis Finnin*)

smelly ones. Eventually, rains and winds wash away anything that remains, and no sign is left that the animal ever lived. A large carcass—that of a bear, say—won't last much longer than a couple of months; even the big bones and the teeth won't last much beyond a season out in the open. Temperature changes from day to night, or from winter to summer, crack apart bones the way they crack bricks and concrete driveways. Nature is an efficient recycler. Dead animals, whether in the Jurassic or in your backyard today, are turned back into soil, plants, and other animals—which die and contribute their substance in turn to the next generation. It's entirely possible that our bodies contain atoms that once belonged to dinosaurs. Arranged somewhat differently, of course.

Occasionally, under certain conditions, decomposition can be staved off for a while. In Yellowstone National Park, it's not uncommon for the spring thaw to uncover the remains of bears that died just as winter was setting in. Buried in snow, they stay more or less intact, like a steak in your freezer, but once the snow melts and the body is gently warmed by the spring sunlight, nature begins to run its course. To hasten the process and reduce the unpleasantness for park visitors, Yellowstone rangers have taken to blowing up dead bears with dynamite. A carcass shattered in this way disappears in about three days.

The same scene could have played out 100 million years ago, with a dinosaur taking the place of the bear—and without the park rangers and their dynamite, of course. Many a dinosaur must have died from disease or in battle with another dinosaur looking for dinner. Occasionally, one would have been lucky enough to die of old age. A dinosaur that died somewhere on an open plain or in a teeming forest would, like a Yellowstone bear, disappear very quickly.

But is there a way that a dead dinosaur could have stuck around longer, without decaying? Let's go back to *Jurassic Park*, to the scene in which Alan Grant, Ellie Sattler, Ian Malcolm, and the hapless lawyer from the Ingen Corporation get to see for the first time the live dinosaurs that John Hammond and his team of scientists have re-created. Gaping from their Jeep, they are treated to a bucolic scene of huge, placid apatosauruses grazing blithely among the trees. (These beasts used to be called brontosaurs, until it was realized that they were identical to the animals earlier named *Apatosaurus*.) But later, when the visitors are trapped in the park with the security fences turned off, it becomes apparent that not all dinosaurs are peaceful, storybook creatures and that Jurassic Park is not an amusement park. This is a real ecosystem, and like all ecosystems it includes its fair share of violence and death. The fearsome velociraptors are fast, aggressive, quick-thinking predators, working in packs to take down much bigger creatures. Procompsognathids—compys, as the rangers have taken to calling them—are the scavengers of the park, the hyenas of the dinosaur world, cleaning up what the raptors leave behind. Muldoon, the former African gamekeeper, coming across a Jeep that's been thrown into a tree by a tyrannosaur and finding no sign of its former occupants, remarks how efficiently predators in the wild clean up, and how little evidence of their feeding they generally leave behind.

Think of a pack of velociraptors attacking one of the docile plant-eating dinosaurs—a maiasaur, for instance. Like lions going after antelope in Africa today, the raptors might well attack an older member of the maiasaur herd or one that strayed away from the pack. If the attack happens out on the open plain, and the raptors eat what they want and leave the rest behind for smaller scavengers to clean up, pretty soon nothing will be left.

But suppose the raptors attack a maiasaur grazing in a swamp or in the shallow water at the edge of a lake, and imagine that the bigger creature rolls over on some of the raptors as it struggles and dies. The maiasaur and some of the raptors drown, and their bodies sink into the mire. Like the Yellowstone bear buried in snow for the winter, the remains of these dinosaurs last longer in the deep muck at the bottom of a swamp than they would out in the open. But how much longer?

In September 1991, two hikers in the Austrian Alps came across what they first thought was the partially preserved body of another hiker, perhaps one of the handful of people who disappear without trace in the mountains every year. But it was an oddly dressed hiker, wrapped in rough furs and wearing what appeared to be laced sandals of animal hide. The true identity of this ancient mountaineer will never be known, because forensic investigation by scientists from all over Europe has revealed that he lost his way about 5,000 years ago. Perhaps he drowned in a cold stream and was carried into some quiet nook, where he lay frozen over the centuries. Thousands of years later, the slow movement of the glacier he was trapped in brought him out into the open again. It's not unusual for the bodies of lost climbers to stay hidden in glacier ice for long periods of time: just a few weeks before the discovery of the unfortunate iceman, the remains of two Alpine hikers missing since 1934 were found.

Oetzi, as the iceman was named (from Oetztal, the name of the mountain valley where he was discovered), may have died in an unlucky accident, but the almost perfect preservation of his frozen body was a piece of luck for modern scientists, who were able to determine not only what kind of animal skin he was wearing but what he had eaten just before he died. More than that, scientists were able to extract a few bits of his DNA, which turned out to be nearly identical to the DNA you would extract from any modern inhabitant of northern Europe yet different in small but significant ways from the DNA of South American Indians or sub-Saharan Africans. Oetzi really was a European.

Even older than Oetzi is a deep-frozen woolly mammoth that was found in Siberian ice some years earlier. This creature is reckoned to be about 40,000 years old, and its skin, flesh, and fur were remarkably well preserved. Mammoths stomped the earth long before Oetzi went on his fatal hike in the Alps, and they became extinct in the Pleistocene. Recently, a Japanese scientist announced that it might be possible to resurrect the mammoths by extracting frozen sperm from mammoth remains and using it to inseminate an elephant. Successive breeding with preserved mammoth sperm, he speculated, should create a series of increasingly mammothlike animals. This idea has been met with considerable skepticism; it's conceivable in the first place only because frozen mammoth remains are relatively young by dinosaur standards. Compared to the vintage of the dinosaurs, 40,000 years is a fraction of a second.

Here and there around the world, conditions have allowed human or animal remains to last for an unusually long time. A cave called Ultima Esperanza ("Last Hope"), in the desert of southern Chile, produced a 13,000-year-old sloth that had been completely dried out by the wind, like a big piece of beef jerky. In a peat marsh in Cheshire, England, a 2,000-year-old human body, leathery but recognizable, was found in 1983. This character—dubbed Pete Marsh, by the radiographers at a local hospital who were examining him and needed a name for their paperwork—was preserved because the heavy muck of the bog had kept out oxygen, which promotes decay, and the acidity of the bog tended to kill bacteria and effect chemical changes in the flesh which made it less prone to decomposition. Pete Marsh was pickled, in other words.

The preserved remains of similarly primitive Europeans— hundreds of them, in fact—have been found in the bogs of northern Europe. In 1996, a "Peruvian princess," the 500-year-old body of a young girl, was found frozen high in the Andes. The La Brea tar pits of Los Angeles have trapped and preserved a quantity of rodents, wolves, saber-toothed tigers, mastodons, and other animals that stumbled into the sticky mess about 10,000 years ago. And Egyptian mummies, preserved by design rather than accident, have lasted several thousand years. The oldest fragment of preserved human DNA comes from 8,000-year-old human remains found in Windover Pond, an acidic bog in Florida that was apparently used as a burial place by early Floridians.

All very interesting, you're thinking, but what does this have to do with our buried maiasaur and its raptor attackers? Unfortunately, not very much. Though an occasional dinosaur may have got itself pickled in a peat bog 100 million years ago, the peat bog itself wouldn't have lasted to the present day. Even over a period of only thousands of years, the earth's surface is a changeable and unpredictable place: peat bogs dry out or are washed away. Over millions of years, they can be buried, or drowned beneath the sea. For the remains of a dinosaur to have lasted to the present day, another sort of process has to occur. That process is mineralization, and it is mineralization that creates fossils.

Bones, the body parts most resistant to decomposition, can survive for a long time in the heavy, airless mud at the bottom of a swamp, long after all the soft tissues have degraded. And sometimes, when conditions are right, the squishy sediment containing the bones is increasingly compacted by the weight of further sediment pressing down upon it. Over the course of millions of years, sediments are compressed into sedimentary rock—chalk, limestone, sandstone, and the like. The same slow changes that turn sediment into rock can turn the bones within the sediment into petrified, rocklike remains—fossils. This happens so gradually that the structure of the original bone is retained.

As water trickles through the sludge and the bones within it, anything that can be dissolved, no matter how slowly, is replaced by insoluble minerals. Because bones are porous, water seeps through them, leaching out the organic content and filling the spaces with mineral material. The trickling water carries atoms of silicon and calcium, which replace atoms such as hydrogen, nitrogen, and carbon in the organic material. The process is unimaginably slow, taking thousands of years to be completed; the replacement of organic by inorganic material proceeds almost an atom at a time—so gently that one atom can take the place of another without any disturbance to the whole. The structure of the original organic material is therefore preserved, but its chemical nature is completely changed. What used to be bone has become, quite literally, stone—a stone in the shape of the dinosaur bone it once was. Fossilized bones can survive for millions of years because all the organic material has been replaced by minerals.

But that also explains why you're not likely to find any DNA in a fossil: DNA is part of the organic content that was replaced by rock. It's gone—broken down and dissolved slowly into the swamp as the fossil formed, and ultimately recycled by nature into new plants and animals. Those atoms may once have formed a dinosaur's genetic code, but their provenance is completely obliterated.

There's one possible loophole in all this. Fossils form from the outside in, and a particularly big bone might be fossilized in such a way that an impervious rocky shell forms around an interior that hasn't been completely leached away and mineralized. That argument came up in 1994, when a group of biologists from Brigham Young University in Utah claimed that they had done what Lewis Dodgson imagined and what we have just about convinced ourselves is impossible: they claimed they had extracted dinosaur DNA from a fossil.

In an account published in the journal *Science*, Scott Woodward and his team reported that they had removed fragments of dinosaur bones—the species could not be identified, but the bones were estimated to be about 80 million years old—from a layer of stone lying above a deep coal seam. (Coal itself is the partly fossilized remnant of plant and vegetable matter that was compressed beneath overlying layers of sediment hundreds of millions of years ago.) The impervious nature of the stone, they believed, had protected the bones from complete fossilization, allowing a little of the organic material to survive. In ultrasterile laboratory conditions, they removed some DNA from this bone and concluded that it was not the DNA of any modern species. It appeared to be as different from mammalian DNA as it was from the DNA of modern birds and reptiles. Although that didn't exactly prove that it was dinosaur DNA, it seemed like a pretty good circumstantial argument, as the lawyers would say. If the DNA came from a dinosaur bone and looked unlike DNA from any modern creature, then the obvious conclusion was that it must be genuine dinosaur DNA.

But Woodward's discovery was greeted skeptically from the outset. Many scientists believed that it was next to impossible to avoid contamination with DNA from other sources. Household dust is packed with dried skin flakes and bits of dead insects, not to mention live insects such as dust mites, all of which contain DNA. Your lungs and your intestines are populated by all manner of bacteria and viruses, which have their own genetic material too. Woodward's case collapsed when several other groups of scientists looked more closely at the identity of the "dinosaur" DNA and discovered that it was of a type that could, after all, be found in people. Although Woodward still argues that he obtained a genuine piece of dinosaur DNA, most scientists now believe that despite all his precautions, tiny snatches of human DNA got into the sample, and that what Woodward and his colleagues extracted, for all their pains, was a more or less inconsequential piece of their own genetic code.

If we need a dependable way to find dinosaur DNA—and enough of it to make several different species of dinosaurs—we had better abandon Lewis Dodgson's idea and look again at what the scientists of Jurassic Park did.

et's return to our primeval swamp, with the maiasaur calmly grazing. Instead of a pack of vicious velociraptors, let's conjure up a swarm of smaller, peskier, more familiar predators. Mosquitoes and other biting insects have been around on earth for a long, long time, and may well have bothered dinosaurs as much as their descendants bother grazing animals and sunbathers today.

Imagine a mosquito biting a dinosaur and drinking a little of its blood. . . .

Hold on a minute—could a mosquito really pierce a dinosaur's skin? The visitors' center on Isla Nublar features a charming little cartoon that takes the viewer through the basics of dinosaur reconstruction. A cartoon mosquito plunges its proboscis into the back of a cartoon dinosaur that looks as much like the real thing as Barney, the purple public-television dinosaur—and about as tough-skinned. Real dinosaurs had thick, leathery hides or (perhaps) plated, scaly armor. Occasionally, fossilization has preserved not only bones but also fragments of this tough skin and the imprint of dinosaur hide on thick mud. It's not too far off the mark to imagine dinosaurs as having something like alligator hide—which is tough enough to withstand mosquitoes. But even so, a dinosaur wouldn't have been completely impregnable. No matter how tough its hide, there must have been softer, more flexible skin around its eyes and mouth—think of the way flies congregate around the eyes of a horse or a cow. And if the dinosaur was protected by rigid, scaly plates, there would have been joints and gaps between the plates so that the animal could move its limbs. Where there's a gap, an insect could find its way in. Paleontologists have found ancient insects with fearsome equipment on their mouth parts—equipment that seems designed for piercing thick skin. We can never be sure that such insects actually fed on dinosaurs, but they may have. And these insects would thereby take on board a sample of the dinosaur's blood and some DNA along with it.

Let's pick up the scene again. After taking a sip of dinosaur blood, the mosquito flies off to rest on a tree and digest its meal. But it happens to land on a tree that's oozing gum from a wound in its bark; perhaps a passing triceratops has scraped its horns against the tree to clean them off. The insect's feet are trapped in the sticky gum, and as the gum continues to flow, it rolls over the luckless mosquito, covering it completely. Exposed to air and sunlight, the gum hardens, and after a matter of hours or days it has solidified into a resinous mass with the look and feel of tough, yellow plastic.

It's largely an accident of nature that amber happens to be such a remarkable preservative. Tree resin is basically medicinal: it hardens over damaged bark to protect the tree as it heals, and it contains antibacterial chemicals that keep decay and disease at bay. The antibacterial qualities of the resin also help to preserve the corpses of trapped insects. Moreover, amber is a powerful desiccant—it soaks up water molecules and locks them away within its chemical structure. Egyptian priests used to seal mummified pharaohs with pieces of amber as a way of keeping the pharaohs' remains dry. (That quality happens to be particularly useful in preserving DNA, which tends to break up in the presence of liquid water.)

These properties of amber, along with its incredible durability, explain why it's an excellent preserver of insects. But from the tree's point of view, the fact that fully hardened amber can survive for millions of years seems a little excessive. After all, the tree itself isn't likely to be around for more than a century or two, yet the Band-Aid it manufactures can linger for eons—a situation roughly analogous to equipping a heart pacemaker with a 10,000-year battery. And, in fact, only a few kinds of tree resin harden into amber that will last millions of years. Jurassic pine trees probably generated little droplets of sticky yellow resin as their modern descendants do, but pine resin doesn't generally become tough enough to survive for thousands of years, let alone millions. Other tree species produce gums or resins containing sugary substances that bacteria feed on, so these are decomposed before they have a chance to harden. And other resins harden, but not completely. The true ambers harden completely, their chemical nature changing ever so slightly over the decades to become steadily tougher and more impermeable.

But even these ambers have to be protected. If they're lying around on the ground, they may be pulverized by the tread of a dinosaur. And in air, they will slowly absorb oxygen, which prevents some of the toughening chemical changes and allows them to weaken and, eventually, crumble. Even the most durable ambers, if they're to last a million years or more, have to be buried somehow. One likely scenario is that when trees carrying their amber jewels die (or are knocked over by a lumbering apatosaurus) and fall to swampy ground, they embed their impervious fragments of amber in heavy muck. In the Dominican Republic, pieces of amber are trapped within buried wood that has turned to lignite, a kind of coarse coal. But as we discovered in the prologue, Dominican amber is not old enough for our purposes. To find a mosquito that might have bitten a dinosaur, we need to go somewhere else—New Jersey will do fine, although parts of northern Canada will fill the bill, too. If you insist on an exotic location, we can take Spielberg's film crew to Lebanon, where there is amber that's 125 million years old and happens to contain quite a few species of insects. The trees that produced the New Jersey and Lebanon amber are mostly either cypresses or varieties of *Araucaria* and *Hamamelidaceae*.

At any rate, we don't really care where the amber came from, as long as it hardens into chunks that can survive for at least 65 million years. Sifting through this amber is no easy feat. Finding it in the first place takes a certain amount of expertise. In the New Jersey field, you have to sort through the heavy clay soil by hand, feeling for lumps, most of which will turn out to be ordinary gravel. As noted, a piece of amber in the ground won't look very remarkable—it will be dirty, coarse, scratched, and dull. The early scene in *Jurassic Park* in which a

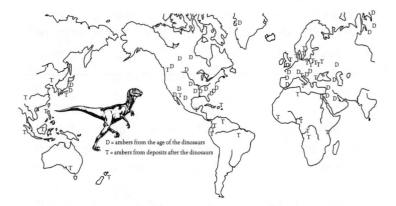

D = ambers from the age of the dinosaurs
T = ambers from deposits after the dinosaurs

Amber fossils are found on all continents except Antarctica. The *T*'s indicate sites where Tertiary ambers and copals are found. These ambers range in age from very recent to 45–50 million years old. The *D*'s indicate sites where amber deposits from the time of the dinosaurs exist. These deposits range in age upwards from 85 million years old. Most are from the Cretaceous period, but a few range back as far as the Jurassic and the Triassic—although the oldest amber is generally less well preserved.

miner holds a nugget of amber up to the light to show off the big bug it contains is a touch of *cinéma faux*. In most cases, it's only when you get the amber back to the museum, where you can clean it up and polish it, that you can start looking for insects.

The mining and collection of amber is quite an industry in several places around the world, so *Jurassic Park*'s John Hammond doesn't have to get his hands dirty by digging holes in the ground himself. He first comes to the attention of United States government officials because he has spent $17 million stockpiling amber from sources and traders all over the world. He even buys up amber jewelry: amber ornaments with insects inside are a popular novelty item.

Scientific amber collectors would never buy on the commercial market like this, because you have no way of knowing how old a piece of amber is if it's sitting in a dealer's display cabinet. To know its age, you have to know what kind of rock or soil it came from. Dealers are interested only in the amber's appearance, not its age—and in how much they can sell it for. They may lie about the origin of the pieces they're selling, if the lie will bring in an extra dollar. There are even forgeries: people will drill a hole in a genuinely old amber nugget to insert an altogether modern insect. Recently, insects embedded in synthetic polyester resins have been offered to amateur collectors as the real thing. The careful scientist will visit amber mines in person, to be sure of the authenticity of the amber. John Hammond's problem, however, is that he needs all the amber he can get to search for the handful of pieces that contain what he's looking for. Even if you find a piece of amber with an insect in it, the chances that it has dinosaur blood inside it are small, so Hammond wants, in effect, to corner the market.

So we are left with New Jersey, Lebanon, Canada, and a few

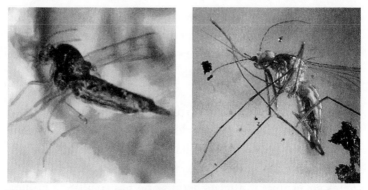

The insect on the right is a mosquito from Dominican amber and the one on the left is a female no-see-um from New Jersey amber. Both have biting or piercing mouthparts, but only the one on the left is old enough to have possibly bitten a dinosaur.

(*Photos by Dr. David Grimaldi*)

scattered sites in Europe and Asia. Let's assume that among all the amber Hammond manages to buy there's a good deal that could be old enough to contain insects that might have bitten dinosaurs. Now it's just a matter of slowly, carefully, painstakingly sorting through it all.

Paleontologists have found plenty of plant fragments in amber—bits of grass or leaves, seeds and grains, even a whole mushroom, and a flower cluster from an oak tree—but these items aren't of much interest to dinosaur hunters (although Hammond, early in the book, asks Alan Grant for details of what dinosaurs eat, and the existence of amber-preserved plants is one way to know what sort of vegetation dinosaurs would have grazed on). More spectacularly, there have been discoveries of spiders, scorpions, even whole frogs and lizards in amber.

But what about insects? New Jersey amber is rich in insects, the great majority of them being gnats and midges. A few of these are biting insects: you can find midges pretty much like the typical "no-see-ums" that still infuriate the modern inhabitants of the area. There's the 85-million-year-old mosquito recently discovered at the American Museum of Natural History in New Jersey amber. And a nasty bloodsucker called a sandfly has been identified in a piece of Lebanese amber 125 million years old.

On the other hand, no one has yet found a preserved insect from the age of the dinosaurs with its stomach full of fresh blood. Even if a blood-filled insect were found, it's hard to guess what state of preservation the blood might be in. Nevertheless, the first part of the *Jurassic Park* problem—finding a suitably old insect with dinosaur blood inside it—doesn't seem impossible. As David Grimaldi, an AMNH entomologist who specializes in fossil insects, remarked recently, "If one were searching for dinosaur blood, an amberized fossil insect would be an imaginative place to look for it—but then the real difficulties would arise."

CHAPTER TWO

BLOOD FROM A STONE

You're in your laboratory, examining a piece of amber about the size of a walnut. If you hold the amber up to the light, you can plainly see, inside it, a mosquito. This mosquito is 85 million years old. You don't know that it ever drank blood from a dinosaur, but it might have. And you certainly don't know that it drank dinosaur blood just before it got stuck in amber—but it might have.

Now what? You have to get the insect out of the amber or at least extract the contents of the insect's stomach. Do you shatter the amber with a hammer? Saw it in half? Dissolve it in chemical solvent? Melt it?

In the orientation film at Jurassic Park's visitors' center, we see a little hole being drilled into the amber surrounding an ancient mosquito. A needle is inserted, and a syringe draws out something—we're not sure what yet—from the insect's stomach. Easy! But maybe not such a good idea.

Any piece of amber, no matter how pure and translucent it looks, contains all kinds of microscopic bits and pieces of the past: pollen, midges, nematode worms, fragments of twigs and petals. All these things have their own DNA, and if you drill a hole through the amber and insert a needle, you're going to pick up DNA from any number of life-forms that have nothing to do with dinosaurs. One of the most difficult parts of your task is going to be making sure you have pieces of dinosaur DNA and not DNA from a Cretaceous flowering plant or worm, or even from you. Not to mention that the insect itself has its own DNA, and if you drill a hole to get to the insect's stomach, you can't help going through some of the insect's skin and muscle.

Sawing the amber in half is a better idea, but you have to plan the job carefully. Once that amber casing is cracked apart, the insect is exposed to the modern world and to all the bits of DNA floating around in the air: DNA from viruses and bacteria; DNA from skin flakes and insects; maybe some tuna DNA from that tuna sandwich you had for lunch (there's a stray flake of tuna on your shirtsleeve).

Working slowly, you cut around the circumference of the nugget of amber, using a small circular saw on the end of a jeweler's drill—the kind of tool used by people who make amber ornaments, in fact. You're careful to stop short of the mosquito itself. You want to cut just far enough to allow you to break the amber casing open but not so far that your saw cuts into the mosquito—that would let in all sorts of contamination. At this point, you can wash the nugget liberally with alcohol to get rid of any dust and grime, and especially DNA-containing material from whatever else was in the amber.

Now you're ready to crack the amber open. But you need to do this in a place where all DNA contamination has been thoroughly eliminated, and where any DNA inside the insect will be safe.

Here's how it's done. Ultraviolet light messes up the structure of DNA, altering the form of some of the DNA molecule's chemical components. (That's why UV in sunlight is bad for the skin: it can alter DNA structure in your skin cells and trigger cancer.) Shining ultraviolet light on a potentially contaminated object effectively cleanses it of DNA. In your laboratory, you have a clear Plexiglas glove box equipped with an ultraviolet light. You put all the equipment you will be working with— instruments, chemicals, devices—into this box, and switch the light on. That takes care of any DNA in there. (The Plexiglas is opaque to UV, so nothing outside the box is affected.) Now you put your almost cracked piece of amber in the box, but you've switched the ultraviolet light off because you don't want to destroy anything once you open the amber.

The other feature of this box is positive airflow. Clean, filtered air is forced into the box at slightly more than atmospheric pressure, so that every time you open the box to put something in or take something out, air flows outward. You don't want laboratory air, with all its dust motes and stray DNA, entering the box after you've sterilized it with the ultraviolet light. Even so, you open and close the box as little as possible, and you work inside it by putting your hands into the gloves, which are sealed to the wall of the box. That way, you can manipulate what's inside the box without opening the box itself.

In fact, the whole lab is kept as clean as possible. Ideally, it will be separate from your preparation and storage rooms—and separate, too, from any other labs where your colleagues are working on different types of DNA. You'd have your own set of lab equipment, for use in this work only and not to be mixed up with anyone else's.

Now you're ready to go. First, plunge the amber into supercold liquid nitrogen, which makes it more brittle still. Now, with gloved hands, or perhaps using some kind of surgical forceps so as to reduce the risk of tearing your gloves, you crack the amber gently open, releasing the mosquito from its 85 million years of imprisonment. This will be a nervous moment. As long as the mosquito was in its amber envelope, it was protected from the outside world, but now it's exposed again, vulnerable to contamination. And all those processes of decay and decomposition can start ticking again, after their 85-million-year hiatus.

Your goal, of course, is to get dinosaur DNA, but at this point you don't know if there's any such thing in this particular mosquito. You want to extract as much of the insect as you can to study it further, but that's not so easy. Amber is tough, the mosquito is fragile, and its legs and wings and head are too thoroughly embedded in the amber for you to be able to free it intact. You're going to have to break apart this perfectly preserved creature, but at least you can try, with the aid of a low-powered binocular microscope, to dismantle it carefully, bit by bit. With your instruments you pluck out some muscle tissue and quickly drop it into a little test tube, which you immediately seal up. You have made sure that there's a plentiful supply of sterilized test tubes on hand, and you may well have stuck numbered labels on them all. That way, as you're dropping bits of insect into the tubes, you can call out the number and describe what you're putting in there, so that a lab assistant can write it all down. Many otherwise careful experiments have gone wrong because somewhere along the line someone lost track of which little test tube contained which sample.

So you pull off a leg, put it in a test tube, call out the number. Then a bit of wing, a fragment of skin, some hairs. . . . If you were wise, you would have practiced pulling apart a modern mosquito under similar conditions, to get an idea of how best to do it, and what problems to expect. As with carving a turkey, practice makes perfect.

And you do your best to scoop out the stomach contents, keeping them as separate from the rest of the mosquito as you can manage.

Once the test tubes are tightly sealed to make sure that no more DNA or any other organic material gets into them, you can open the plastic box and throw all your instruments in the sink to be cleaned. Although the immediate danger of contamination is past, you will still go out of your way to keep the test tubes in a secure, uncontaminated place.

Now you have a chance to look closely at the insect parts you've so carefully extracted from their tomb. The exactitude of preservation can be quite astonishing. Even in pieces, the thing looks like a real insect, down to details of the digestive system and the muscles it once used to fly—features you need a microscope to see. It's also possible to make out some individual cells in the various organs. Though amber is a desiccant, the cells of insects that have been trapped in amber don't shrivel up and dry out, as they would have if desiccation were the only way they were preserved. What probably happens is that when the insect is first caught in the sticky resin, some of the lighter and more fluid chemical ingredients of the resin soak quickly into its body, in effect embalming the tissues, just as organs or small creatures soaked in formaldehyde are pickled so that they don't decay. This process of "ambalming," though its chemistry is far from understood, is undoubtedly the principal reason that amber-trapped insects look as if they had been caught and killed in an instant and plunged into a state of suspended animation since then.

Well, now what? You have little pieces of 85-million-year-old mosquito in a series of sealed and labeled test tubes. For the moment, let's forget about dinosaurs and just think about how to get mosquito DNA from the samples. That has actually been done, and learning how to accomplish it will get us ready for the more difficult task of extracting dinosaur DNA—more difficult because it's going to be hard to figure out which insects might have dino DNA in them in the first place. Even if some of the insects do, all we'll find is a minuscule remnant of a dinosaur. If we can't get fragments of insect DNA out of the almost complete body of a well-preserved insect, we might as well give up and go home.

Amber preserves tissues extraordinarily well. The top left photo shows the beautiful preservation of external structures in amber-preserved insects, such as the wing veins in this termite from Dominican amber. The top right photo is a scanning electron micrograph and shows the excellent degree of preservation of muscle tissue in the thorax of a stingless bee from Dominican amber. The bottom left photo is a transmission electron micrograph of thin-sectioned muscle tissue and displays exquisite preservation of the fibrous filaments in this muscle tissue. The bottom right photo is a higher magnification of the third and shows the fine structure of the muscle tissue.

(Photos by Dr. David Grimaldi)

The scientists who study insects trapped in amber are, of course, interested first and foremost in the insects themselves. There's a debate over whether it's ethical or wise to try pulling DNA from amber-preserved insects, because any method you use is inevitably going to break open the perfect seal that has kept the insect in good shape all these millions of years. To get any DNA, you must irreparably damage its unique source.

But you don't care about any of that. You just want some DNA, even if it means blenderizing this marvelous insect into a little puddle of goop. As a matter of fact, that's more or less how you would start.

Again, you're going to be working in an environment that's as clean and sterile as you can make it. Standardized conditions for working with DNA were developed in the 1970s, when this sort of science was first being done, although the reason was mainly to assuage the public's fears about contaminating the environment with altered genetic material. But the principle works both ways: if you set up a barrier between your work and the outside world, you're keeping your samples uncontaminated while you're keeping the enviroment clean.

You've now moved to another lab, away from the place where you were cutting up amber and plucking out mosquito bits. You don't want whatever pieces of DNA you extract getting back into the other room, where it might contaminate the next amberized insect you start to investigate. Your DNA-extraction lab is well away from other laboratories in your department, and has separate entrances and air supply and so on. The lab has plenty of ultraviolet lights to disrupt DNA, and they're kept on all the time—except, of course, when you're working on your insect samples. You don't want to zap the DNA you're trying to extract, let alone blind yourself or give yourself skin cancer. You work at a surgical level of cleanliness, maybe more, wearing clean clothing, perhaps overalls, and a mask with a respirator. The room you work in has a positive airflow, to flush out any contaminants and prevent anything from the outside floating in on the breeze, and you work in a glove box similar to the one you used when you were breaking the amber apart.

In any one operation, you'll be using just a fraction of the insect remains you've recovered, in order not to lose the whole lot if something goes wrong. The DNA you want will be inside cells, so you need to break the cells open. You can do this with a solution of something that's essentially a kind of soap or detergent. Cell membranes are made of fatty molecules called lipids, all bound together, and you use soap to disrupt the membrane just as soap dissolves grease spots on your clothes or dishes. You also add an enzyme that degrades proteins in the resulting mess, because some of those proteins can chew up DNA. Normally, the structure of a functioning cell keeps DNA away from these dangerous molecules, but once you've mashed the cell up, you need to move quickly to keep the DNA out of harm's way.

The mess you've created at this point contains all kinds of stuff besides DNA. The next step is to add phenol, an organic solvent, to the mixture. Phenol is not altogether a nice thing to use; it's smelly, burns the skin, and can cause liver cancer. What this nasty substance does is separate the DNA from other stuff in the mixture. The DNA molecules will end up floating around in the water from the soapy solution, while the phenol will dissolve most of the fatty molecules from the cell structure. The proteins and protein fragments, which don't dissolve in either water or phenol, will form a cloudy layer between the two. To extract as much DNA from the sample as possible, you shake the test tube for a few minutes, so that the contents are thoroughly mixed. After this thorough mixing, you spin the test tube in a centrifuge for a while, which makes the water and the phenol separate like oil and vinegar in a salad dressing that's been left to stand. (The separation would occur if you let the mixture stand, but spinning helps it along.) On top is the water, containing DNA and probably a few protein bits and pieces; then comes the milky layer containing most of the proteins; and below that is the layer of phenol, containing all the other cellular material and junk you don't want. Right at the bottom, you'll probably find remnants of hardier stuff—fragments of insect skin and insect hair, for example.

Now it's a fairly simple matter to suck the water layer from the top with a pipette, which is a fine plastic tube attached to a hand-held device that applies suction. You operate this little vacuum cleaner by pressing a button with your thumb, and it sucks up a tiny drop of liquid. Carefully, you remove as much of the water layer as you can without sucking up any of the milky layer or the phenol layer beneath it. So now you have, you hope, a drop of water containing DNA and not much else. (As we noted in chapter one, DNA tends to fall apart in liquid water. But as long as you don't let the DNA sit in the water for too long, everything's OK. When scientists want to store DNA samples for later work, they generally freeze them.)

This method of separating out DNA is reasonably efficient but (you won't be surprised to hear) not perfect. A few odds and ends of cell membrane will have got into the water layer along with the DNA, so there's one more purification step to be done. In a realistic scene in *Jurassic Park*, a scientist in Hammond's laboratory has just performed this final procedure and is showing off to the visitors a test tube with a weirdly glowing red fluorescent ring. The test tube contains a thick slurry of cesium chloride, which is closely related to sodium chloride, or plain old table salt, but is quite a bit heavier and denser. You make this slurry by taking a couple of teaspoonsful of cesium chloride crystals and dissolving them in the smallest quantity of water you can manage. You fill a test tube with the salty slop, add the water solution that contains (you hope) the insect DNA, and spin the whole thing around very fast in a device called an ultracentrifuge. This machine spins the test tube on the end of an arm going around as fast as 100,000 times a minute—a souped-up, miniaturized tilt-a-whirl.

Rapid spinning makes the cesium chloride solution denser at the bottom of the test tube than at the top. If this is all done right—the right amount of cesium chloride and water, the right ultracentrifuge speed—then the DNA will gravitate to the level at which its density matches the density of cesium chloride at that point. You let the ultracentrifuge slow very gradually, so as not to disturb the contents of the test tube, and you take the tube out carefully. If all has gone well, the DNA will have settled into a single layer in the column of cesium chloride, and all the impurities will be somewhere else.

Before adding the insect DNA to the test tube of cesium chloride slurry, you also added a dye that's designed to chemically attach to molecules of DNA. After centrifuging, you shine an ultraviolet light on the test tube, and you should see a red glow marking the layer where the DNA has ended up with its attached fluorescent dye. Because UV light is damaging to DNA, as we learned a while back, you illuminate the test tube as briefly as possible with a low-intensity light—just enough so that you can mark the position of the DNA layer. You can extract the DNA by sticking a fine needle through the side of the test tube—you remembered to use a plastic one, didn't you? (If you'd used a glass tube, the stress caused by the centrifuge would have smashed it to pieces anyway.)

In the movie, the technician who proudly holds up the test tube with the glowing red ring—revealing, we are meant to believe, dinosaur DNA—is played by a real-life biologist whom Steven Spielberg met when he went to UCLA to get some ideas on what people really do in modern biology labs and what such places look like. As a reward for their help, a couple of the lab folks got bit parts in the movie. Despite their presence, however, there are a couple of problems with the scene. You need a lot of DNA to get a red ring as conspicuous as this one is. That's all very well if you're extracting billions of DNA fragments from a good-sized chunk of a living animal, but not if all you've got to work with are a few dinosaur cells in an insect's stomach.

And then there's a historical consideration: the technology of ultracentrifuging test tubes of cesium chloride was on its way out when Michael Crichton wrote *Jurassic Park* and was defunct by the time the movie came out. The last time anyone at the American Museum of Natural History used an ultracentrifuge for DNA extraction was in April 1990.

Luckily, the second problem takes care of the first. There's a much better and more sensitive way to identify and extract DNA from scanty samples. It's also quite a bit easier.

Let's go back to the point where you separated the layer of water containing the DNA from the phenol layer that has all the stuff you don't want. To the water that holds the DNA, now in its own test tube, you add some ethanol (the official name for plain old alcohol) along with a little salt. DNA molecules will precipitate out in a mixture of water and ethanol, so when you add ethanol you should see the liquid take on a cloudy, white appearance. That white precipitate is DNA.

Next, you put your test tube into a standard-issue lab centrifuge (not the superfast ultracentrifuge you needed before) and spin it around for a while, forcing the DNA into the bottom of the test tube. Now you can pour off most of the water-ethanol mixture and leave the rest to evaporate. If there was a lot of DNA in your original sample, it will be visible as a quantity of white material at the bottom of the test tube. If there

The tube on the left holds the water layer of DNA prep after it has been removed from the phenol layer. In the middle tube, ethanol has been added to the DNA-containing water layer, and you can see the DNA starting to precipitate as a cloudy layer between the water layer and the ethanol. After the tube is shaken a few times, the DNA precipitates out of solution and forms a white floating clump, as in the right tube.   (*Photo by Dennis Finnin*)

was only a tiny trace of DNA, however, it won't be visible; the test tube will look empty, but you carry on anyway, on the assumption that there's a smidgin of DNA in there. You add a drop of distilled water to the test tube, swish it around briefly, and go to the next step—a modern, almost magical step.

You plunk the drop of distilled water into another test tube containing a specially prepared brew of chemicals and stick the test tube into a slot in a hotplate controlled by a timer that raises and lowers the temperature every couple of minutes. Then you go away and have a cup of tea or read a magazine. Two or three hours later, you come back to the lab and pull the test tube from the heater. If all has gone well, there should be plenty of DNA in the test tube—millions of times more than there was a few hours ago. Nice trick, huh?

The secret lies in the chemicals you added before you put the test tube into the automated heater. But to find out what those chemicals do, we need to talk a little more about what DNA is. So far, we've said only that it's a large molecule—but since it carries the genetic code for entire animals, it has to be a molecule with a special structure.

DNA is a chain whose links, little chemical units in themselves, come in four types, designated by the letters A, C, T, and G. A DNA molecule is a string of those letters. In humans, the complete DNA molecule—the human genome—is about 3 billion letters long. (Other mammals have similarly long genomes, while simpler creatures, like bacteria, have much smaller strings—perhaps just a few million letters long.) It's the specific sequence of A's, C's, T's, and G's that makes each individual's DNA different from anyone else's. Nevertheless, every human being's DNA has to contain the same basic building instructions, so large fractions of the human DNA molecule are the same for every person on the planet. The variations control such subtle differences as hair color, shape of nose, predisposition to certain diseases, and so forth, but—interestingly—there are many differences that seem to have no significance at all. Much of every person's DNA has no discernible function and can vary significantly from person to person. It's obviously important that all the sections of DNA containing instructions to build your liver, for example, mustn't vary greatly, because your liver has to do a lot of complicated things and even small departures from normality can be life-threatening. But for a section of DNA that doesn't do much, there's no penalty for change, and these regions of human DNA can vary widely from person to person with no obvious consequences.

For the moment, though, all we need to think about is that DNA is a long string made up of four different letters—or bases, to use the scientific term—in varying sequences. In the cells of your body, however, the DNA does not exist as a single chain but as two such chains spiraling tightly around each other to form the famous double helix. There is a crucial relationship between the two strings: where there's an A on one chain, there must be a T on the other; and where there's a C on one chain, there's a G on the other. The rule is, A pairs with T and C pairs with G. Were this not the case, the two strings could not line up correctly to form the double helix.

In all likelihood, the DNA you obtained from the ancient insect in amber will have broken down over the millions of years of its imprisonment into smaller fragments. What you will have is a test tube containing bits and pieces of insect double helices. The machine that multiplies these fragments of insect DNA does so by heating the test tube almost to the boiling point. At that temperature, the two chains or strands of DNA in each double-helical fragment fall away from each other, although the individual strands remain intact. Then the machine lowers the temperature. When the mixture is at about 72°C, the special ingredients you added start to kick in. Those ingredients include a plentiful supply of the individual bases—the A's, C's, T's, and G's—floating freely about. In addition, there's an enzyme called DNA polymerase, whose job is to ratchet along a single strand of DNA, reforming the double helix by fashioning the correct adjacent strand. In other words, if the polymerase is at a T, it will pick up one of the free-floating A's and attach it across from the T; if it's at a C, it will find a G and do the same.

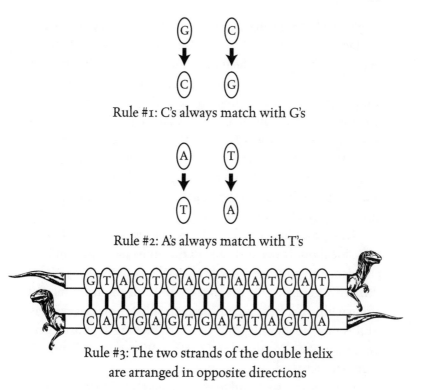

Rule #1: C's always match with G's

Rule #2: A's always match with T's

Rule #3: The two strands of the double helix
are arranged in opposite directions

In constructing double-stranded DNA from the individual bases, A's always pair up with T's, and C's always pair up with G's. In addition, each base has a "head" and a "tail" end, and the two strands run side by side in opposite directions.

Enzymes are the mechanics and janitors of the biochemical world. Each one has a particular job. Want to break down that alcohol you just drank with your beer? There's an enzyme in your liver that will do it. Want to clear your bloodstream of fat from that Thanksgiving dinner you just ate? Other enzymes will take care of that. What the enzyme polymerase does is manufacture the right DNA sequence to pair up with a single DNA strand that's already there. Where there was one double helix to start with, now there are two. And when the machine goes through the temperature cycle again, the process repeats. While you are drinking your cup of tea or reading your magazine, the machine repeats this cycle over and over, doubling the amount of DNA at each step. Five repetitions multiply the quantity of DNA 32 times; another 5 give you 32 squared, which is just over 1,000—1,024, to be precise. Twenty cycles give you a million, and 30 cycles, which the machine can complete in a matter of hours, will multiply the original amount of DNA a billion times.

This procedure is called the *polymerase chain reaction*—or PCR, for short. It was invented in the mid-1980s by a flamboyant biochemist named Kary Mullis, who won the 1993 Nobel Prize in Chemistry for his invention. With the prize money, along with what he's earned commercially for his clever ideas, Mullis now divides his time between working as an independent scientific consultant and surfing. Who says brilliant scientists have to be bench potatoes?

PCR became so useful to researchers that barely 5 years after its invention you could buy from mail-order scientific-equipment catalogues a machine that did the whole thing for you. The simplest machines have metal hotplates with holes that can accommodate 50 or more test tubes at once, and a little computer that you can program to set the number of cycles you want the machine to go through and the rate at which it performs the cycles. A couple of thousand dollars will get you a plain-Jane PCR machine. For $12,000, you can get something a little fancier: some machines have robotic arms that pick up the test tubes and dunk them into tubs of hot and cool water; others blow hot and cool air over the test tubes. PCR efficiency is best when the temperature changes are as fast as you can make them.

One further innovation made it possible for PCR to become as simple and automated as it is today. Although we talked about the enzyme called polymerase as if it were just one thing, there are really many versions of it. Almost every creature or organism has its own, slightly different version of polymerase. A problem with the original PCR technique was that the high temperature needed to break the DNA strands apart was also high enough to destroy the polymerase. This meant that you had to add a new supply of polymerase at every

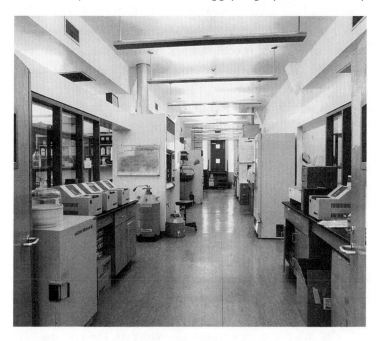

A 180° turn from those dusty dinosaur fossils on page 3 faces you directly into the molecular laboratories at the AMNH. This photo shows the entrance to the molecular laboratories at the AMNH, two floors above the dinosaur halls. Just inside the doors to the lab and to the left sit three of the eight PCR machines that essentially drive the genetic research of the laboratory.

(*Photo by Dennis Finnin*)

step, when the test tube was cooler again—so that PCR required constant tending, making it just one more lab operation that demanded a lot of time and manpower and patience. The innovation—an ingenious one—was to use the version of polymerase from a one-celled organism called *Thermus aquaticus*. This creature, as its Latin name indicates, likes to live in hot water. It was first found in hot springs in Yellowstone Park, and it can survive in water that's almost boiling. Its polymerase—Taq polymerase, for short—can handily withstand the extremely high temperature used in PCR. Taq polymerase becomes inactive at such temperatures but starts working again when the test tube is cooled down. You can use one batch of ingredients through a few dozen cycles without having to add anything new.

This is what allows you to go for your tea break while the PCR machine is humming quietly along. Eventually, you come back to the lab and pluck the test tube out of the machine. If all has gone well, there will be lots of DNA where there was originally just a little.

So now we're all set to go prospecting for dinosaur DNA. With the help of PCR machines, we have a way of reproducing even scanty DNA samples from whatever ancient insect is available. On to the next step.

Well, it's not quite that easy. Polymerase, the enzyme that attaches A's, C's, T's, and G's to single DNA strands to make new double helices, doesn't work unless it's given a kick start. If all you have in your test tube are single DNA strands, Taq polymerase, and free-floating bases, nothing will happen. The polymerase first has to latch onto the single strands, so it can start ratcheting along. And to attach itself, it needs help.

The other ingredient in the chemical mixture that goes into the PCR machine is what's called a "primer." This is a single strand of DNA typically 15 or 20 bases long. After the machine has heated the test tube to break apart the double helices, it cools the solution down to about 50°C—cool enough to allow the primers to hook up to the longer strands of separated DNA, making the beginnings of new double helices. Then the PCR machine brings the temperature back up to about 75°C, which is the temperature Taq polymerase likes. Now it can get to work adding more letters, creating a new double-helical DNA strand, making a copy of the original.

All right, you think, so we need to add primers. What's the big deal?

The big deal is this: a primer will be able to attach itself to another single strand of DNA only if it has the complementary sequence: that is, if the primer sequence (to pick something completely at random) is ACTTGACCTGAAGTT, then the one and only sequence it can hook up to is TGAACTG-GACTTCAA. Recall that there has to be a T opposite an A, a G opposite a C, and so on. If all goes well, the primers will find places *somewhere* along the length of the DNA you're trying to reproduce—somewhere there may well be the correct matching sequence. They'll latch on, the polymerase will have a place to start working, PCR will start multiplying, and you're off to the races. But if there's no sequence that matches the primer, the primer can't latch on, the polymerase can't go to work, and nothing will happen.

Scientists use PCR all the time, whether for multiplying bits of ancient DNA from insects trapped in amber or for attempting to find DNA "fingerprints" in a murder investigation. But in these cases, you already have an idea of what you're looking for. Insect DNA, for instance, contains certain sequences of letters which show up in a wide range of modern insects but don't appear in reptiles or mammals. So you compose primers to match known insect sequences and use them to see if you can amplify DNA from, say, a 30-million-year-old termite. If you know what primers to use, all is well. PCR can amplify tiny bits of DNA from dried blood, ancient insects, skin samples, maybe even saliva from the stamps the Unabomber licked.

But to use PCR to multiply dinosaur DNA, you need primers that will latch onto dino DNA sequences. Except that before you've found any dino DNA, you don't know any of its sequences, so you can't make primers you know will work. If you don't know what you're looking for, how can you find it?

CHAPTER THREE

PICKING UP THE PIECES

You're tagging along with Alan Grant, Ellie Sattler, and Ian Malcolm on their first visit to Jurassic Park. Henry Wu, the enthusiastic chief geneticist lured away from what John Hammond derisively calls "the intellectual backwaters" of university research, is showing you around the lavishly equipped research laboratories of Isla Nublar. Better than anything he could have aspired to in the uncertain, cost-conscious academic world!

Wu is explaining how they get dinosaur DNA. First, they collect bloodsucking insects of the right age, preserved in amber. Second, with a microscope and a fine syringe, they poke into the insects' stomachs to see what they can find. Only a few insects have any dinosaur blood in them, but for each dinosaur species they need only one blood sample. Third, when they've found a suitable blood sample, they carefully draw into the syringe as many red blood cells as they can find. Any one of those little cells contains the entire DNA of the creature it came from. It's easy to be sure that a red blood cell inside the insect doesn't belong to the insect itself, for the simple reason that insects don't have red blood cells. They have a kind of blood, called hemolymph, in which oxygen-carrying molecules float around freely rather than being packaged in blood cells.

Wu admits that the DNA they find this way might not be perfectly preserved but says they've come up with methods to patch together pieces that have fallen apart, and even to fill in occasional gaps. Nevertheless, once they've got hold of a single dinosaur blood cell, they've basically solved the problem. On to the next step, which is turning dino DNA into an actual dinosaur.

Wait! There are a couple of little details we need to think some more about.

First, there's a problem Wu mentions only to explain why it's really not a problem after all. Human red blood cells, as it happens, don't contain a full complement of DNA. Red blood cells live only for about 100 days, and their only purpose is to transport oxygen in the bloodstream. As they die, they are constantly replaced by new ones. Because the body has to keep churning out a steady supply of red blood cells, it may be a matter of biological efficiency that the need to fully equip them with DNA was dispensed with at some point in evolutionary history.

But dinosaurs, Wu wants you to believe, carried the complete genome in their red blood cells. This is not such an unreasonable idea. Some dinosaurs are thought to be the ancestors of birds, and red blood cells from birds indeed contain the entire DNA molecule. The reason that red blood cells in birds and humans are different from each other is that— well, birds and humans are different from each other. No one really knows what dinosaur blood looked like, or how it worked, or what sort of ingredients it had swimming around in it, but Henry Wu is probably on pretty safe ground in thinking that it was more like bird blood than like mammalian blood.

Let's give Wu the benefit of the doubt and suppose that the red blood cells of dinosaurs carried the complete dinosaur genome. But maybe the whole question is irrelevant: even though human red blood cells don't contain the entire complement of DNA, a sample of human blood *will* contain the full genome, simply because there's other stuff in there—white blood cells, for example, which are important in fighting disease. If that's the case, though, then why does Henry Wu want a *red* blood cell from a dinosaur rather than just any old cell he can lay his hands on? A simple answer is that blood contains more red blood cells than anything else, so it's the most likely thing you'd find in a blood sample. And indeed, red blood cells from a dinosaur would be easy to distinguish from any stray insect cells.

So you're back in your lab, looking through a microscope at the insect parts you've carefully extracted from amber. You find the remains of the insect's stomach and poke around until you find something that looks like a complete cell from whatever dinosaur or other animal this insect was feeding on before it got trapped on a sticky Jurassic tree and swallowed up in resin.

Unfortunately, there's another big problem that's been overlooked here.

Why does an insect drink blood? To eat, of course. And what does an insect—or any creature, for that matter—do after it's eaten? Why, it grabs the newspaper, turns on the TV, settles down on the sofa, and digests its dinner.

Ah! Digestion. . . . The blood ends up in the insect's stomach, and that's a tough place, awash with caustic chemicals and digestive enzymes whose very purpose is to mash up and break apart everything that comes along. There are also microorganisms in the insect's digestive system which help break down certain kinds of food. The whole point of eating is to take in complicated structures of proteins, fats, and carbohydrates, and then to pull them all apart so that you can incorporate those necessary substances into your own body. In addition, chemical reactions in the digestive system extract energy from food by breaking apart its molecular structure in particular ways. Given all this, a simple red blood cell plunged into the harsh environment of an insect's digestive system won't survive very long. Once the cell is ruptured, its entire cargo, DNA included, is bathed in a vat of destructive digestive chemicals. To the creature it comes from, DNA is the molecule of life, the carrier of genetic inheritance. To anyone else, it's just a snack.

Of course, we're assuming that the insect we're now looking at didn't survive very long after eating but got stuck in resin, engulfed, and then trapped in amber right after dinner. But that sequence of events takes a while. Even if an insect landed on a resin-coated tree immediately after eating, it would take some time for the resin to engulf and kill the insect, and for the resin's preservative chemicals to seep into the insect's tissues and put a stop to decay.

And after the insect is completely trapped in resin and completely dead, the digestive reactions in its stomach don't instantly come to a halt. The stomach's chemicals continue to do their destructive work. Tree resin takes hours, days—even years, in some cases—to harden into an amber that can last millions of years. But just a few minutes inside an insect's stomach is enough to start wrecking a dinosaur's DNA. It's really not at all likely that an insect could have fed on a dinosaur, become instantly trapped in resin, and then become preserved quickly enough so that some of the dinosaur's blood remained undigested. Remember that all the DNA that's been extracted so far from organisms in amber came from the organisms themselves. The idea of extracting dinosaur DNA from an amber-trapped insect is a huge step beyond what any scientist has currently achieved and doesn't seem like a strategy you could rely on.

At this point, you might consider abandoning the notion

of looking for dinosaur DNA inside an insect and instead try to come up with a more likely possibility. It was *Jurassic Park's* seminal notion that such an apparently straightforward and commonplace occurrence as a mosquito biting a dinosaur could lead, with a bit of luck and ingenuity, to the possible re-creation of live dinosaurs. Sadly, here is a lesson that every scientist learns, and usually pretty quickly: it's easy to have clever ideas, but it's rare to have clever ideas that actually work. The solution is to come up with another solution. You know that bits of DNA have actually been extracted from insects trapped in amber, so the possibility of DNA preservation over tens of millions of years is not a fantasy. If an insect with at least some of its DNA intact can survive for that length of time, why not a piece of dinosaur?

Dinosaurs must have fought, and in an especially vicious fight a few chunks of dinosaur might have flown through the air. Or, since dinosaurs die and their bodies get eaten by scavengers, perhaps some particularly messy eater tearing dinosaur meat apart would have left a few morsels lying around. One way or another, let's imagine that a piece of dinosaur could have been flung into a suitable tree, swallowed up by resin, and then preserved in hardened amber to the present day.

Needless to say, no such thing has yet been found. There's an inch-long section of snakeskin preserved in 25-million-year-old Dominican amber, presumably because the snake shed its old skin near resinous trees. As noted, whole frogs and lizards have been found in amber—although small ones, of course. Is it so unreasonable to think of finding a piece of dinosaur in amber? Any part of the dinosaur will do—a chunk of flesh and muscle, a piece of liver discarded by a fussy scavenger, a sliver of skin sliced off by a sharp rock.

Assume that John Hammond has bought up every last piece of amber in the world, scoured every amber mine, and found what you need: chunks of amber containing chunks of dinosaur. You can't tell directly whether a chunk of meat inside amber came from a dinosaur or a little Jurassic rodent, but that was also a problem with the original scenario: you couldn't be sure that blood inside an insect was dinosaur blood and not rodent blood. Any way you look at it, there's going to be a lot of trial-and-error in this part of the project. You'd have to hunt for DNA in a lot of different samples without being able to figure out until later whether the DNA came from a dinosaur or some other creature.

Now you're back in the lab, looking not at ancient mosquitoes but at morsels of (what you hope is) dinosaur steak trapped in amber. That's progress, but your problems are far from over. Your preserved piece of meat may well have some DNA in it, just as amber-preserved insects and plant fragments are known to have DNA in them. But what's been extracted from insects and plant fragments—so far, anyway—is no more than tiny snippets of DNA, not the animal's complete genetic code.

The first example of DNA archaeology came from the lab of Allan Wilson and Russ Higuchi, at the University of California at Berkeley. In 1983, they extracted DNA fragments from an extinct horselike animal called the quagga, which used to live in South Africa but was hunted to extinction toward the end of the nineteenth century. Wilson and Higuchi took scrapings of dried muscle from a quagga pelt that had been preserved in salt in the Berlin Natural History Museum. This was in the dark ages of molecular biology, before Kary Mullis had invented PCR, and Wilson and Higuchi had to use a much more laborious and difficult technique to get DNA. They mashed up the bit of dried quagga pelt and used standard chemistry (the same sort of phenol-and-water cocktail we used in the last chapter) to separate out any bits of DNA that might be in there. Then they added a dose of a genetically tailored virus designed to incorporate stray bits of DNA into its own genetic machinery. This is a somewhat hit-or-miss procedure, but by controlling the amount of virus you add and the length of time everything is allowed to blend, you hope that every fragment of DNA in the original mixture gets picked up by a virus.

The trick is that these viruses, called bacteriophages, are of a type that will infect many kinds of bacteria. For convenience, the bacteria that biologists often use for this purpose are *Escherichia coli*, a type found everywhere—in water, soil, and in your intestines, where it forms part of a permanent population that helps your digestion. You stir the viruses up with a lot of *E. coli*, and after allowing time for them to infect the bacteria, you spread the *E. coli* out on a plate covered with a nutritious jelly and let the bacteria grow and multiply.

Inside the *E. coli*, the bacteriophages, too, multiply. They have been fine-tuned in such a way that they will kill their host bacteria only if they have picked up a piece of foreign DNA—in this case from the quagga. After you've left the plate of bacteria to grow for a while, you'll see a "lawn" of bacteria covering it, with holes here and there. The holes are where the killed bacteria have literally spilled their guts, throwing out their contents onto the plates. Those contents will include the killer bacteriophages, carrying their load of quagga DNA, which is now present in lots of copies because the phages themselves have been multiplying. Now you have to scrape up the cellular remains from the holes in the lawn and look at the DNA there. We won't even begin to tell you how much work it is to fish out DNA and identify the bits that don't belong either to *E. coli* or the bacteriophage: those unidentified pieces must be quagga DNA.

You can see why PCR became so popular so quickly. The process that Wilson and Higuchi used was painstaking, tedious, and unreliable. It might take a week of fiddling and fussing to get anything. Nevertheless, they did succeed in finding some quagga DNA, and by comparing it to related segments of DNA from modern horses and zebras, they were able to figure out where quaggas belonged in the equine family tree (they were more like zebras than horses, as it turned out).

This was the first successful extraction and identification of DNA from an extinct animal. But none of the DNA segments that Wilson and Higuchi found were more than a few hundred "letters," or bases, long. How long is that? Not very, unfortunately. Recall that the complete DNA for a human being runs to something like 3 billion bases, a figure that's typical of many large animals. A fragment of a few hundred bases is a very small piece indeed. If the complete quagga genome were a phone book containg the names of a million people, such a fragment would be a couple of letters out of one person's name!

On the other hand, the quagga pelt hadn't been preserved in an especially careful manner. Amber should do a better job of preservation than a drafty cabinet in a storage area of the Berlin museum, so maybe you'll have a better chance of getting more and bigger fragments of DNA from amber-preserved remains, even though they're tens of millions of years old instead of just a century.

How well amber can preserve DNA is impossible to predict. DNA is not a particularly robust molecule. In a living organism, safely ensconced in functioning cells, DNA is surrounded and safeguarded by all kinds of protective biochemical machinery that keeps it intact and can even repair some kinds of damage. But once the organism dies, that machinery shuts down, and DNA is just another lone, unprotected molecule. Working in Britain, the biochemist Tomas Lindahl has estimated that a piece of DNA in a watery environment (such as the peat bog that preserved Pete Marsh) will break completely apart into its individual letters, or bases, on a timescale of something like 30,000 years—probably faster than that in the rather acidic environment that Pete Marsh occupied. Over time, the chemical bonds that hold the bases together in sequence tend to fall apart. For that reason, Lindahl is skeptical of any of the scientific claims involving recovery of even short fragments of DNA older than 30,000 years. But while DNA in amber might survive longer than that—because the amber dehydrates whatever's inside it—there is still a question as to whether a substantial fraction of the molecule can survive for millions of years as opposed to a few thousand. The oldest DNA fragment that has ever been recovered was from a 125-million-year-old beetle, and the fragment itself was only 400 bases long. That length is typical of the DNA fragments that have been retrieved from all manner of very old amber-preserved insects and plants.

Before you start analyzing your ancient meat sample, one thing you might try is grinding up one of the whole frogs that have been found in amber, to see what DNA you can get out of it. These little creatures are complete, seem to be exceptionally well preserved, and are a mere 30 million years or so old. If it turns out that the DNA of an entire 30-million-year-old frog captured in amber is highly degraded and fragmented, the chances of your being able to get anything valuable out of a piece of preserved dinosaur 3 or 4 times as old look pretty slim. There's nothing you can do about this part of the problem. You can only hope that nature has been kind.

But that brings us back to the problem we ran into at the end of the last chapter. You can use PCR to search out and multiply individual fragments of DNA, but in the standard version of PCR you have to have some idea of what you're looking for.

You have to supply primers, but if you're looking for DNA of an unknown type, then by definition you don't know what primers to use.

Once again, we're going to have to be inventive.

You've carefully removed the alleged piece of dinosaur from its amber casing, using all the tricks you've learned in order to avoid contamination, and you've dissolved the tissue in the soapy solution that breaks cells apart. You've added phenol to the mixture in order to separate out the DNA. Now you've got a little test tube containing microscopic quantities of what you hope is dinosaur DNA. How do you detect, multiply, and extract the DNA? And how can you be sure that you've retrieved all of it, since every little fragment is precious? Here are a couple of ideas.

One thing to try is a direct adaptation of the methods used to get DNA from ancient insects. There you use PCR primers derived from modern insect DNA in the hope that they will pick up similar DNA sequences in ancient insects. You can't put your hands on any primers derived directly from dinosaurs, but you can make primers from bird DNA. If birds descended from dinosaurs, then it's plausible that some DNA segments in modern birds could resemble segments that once belonged to dinosaurs.

Taking this idea seriously, you could analyze the DNA of a whole variety of modern birds, covering a wide range of different species, with the aim of identifying certain sequences that appear in all birds. You might also look at the DNA of some lizards or snakes or amphibians, which are thought to be more distant dinosaur relatives; anything that's common to birds and lizards may well represent a piece of a more primitive genetic code that might have been shared by dinosaurs. With PCR primers based on these common sequences, it's possible that you could pick up and multiply the sections of DNA in your ancient sample.

On the other hand, figuring out all those bird DNA sequences is a pretty tall order. What's more, this technique would pick up only those DNA sequences that have evolved relatively little over the past 65 million years—the ones that are most closely related to DNA in modern birds and, therefore, least characteristic of the genomes of dinosaurs.

What you really need is a method that can pick up every last piece of DNA from your sample without any second guessing as to what the DNA might look like or where it might have come from. This strategy amounts to doing PCR with "random primers": if you don't know what kinds of primers to use, try all the ones you can think of. Any primer is a piece of DNA, a string made up of the bases A, C, T, and G—and in a typical PCR application the primer will be some 20 bases long. How many such primers can there be, altogether? In each position of the 20-letter sequence, you can put one of the possible 4 letters. Using 2 letters, you'd get $4 \times 4 = 16$ possible pairs of letters; using 3 letters, you'd get $4 \times 16 = 64$ possible triplets. The number of different 20-letter primers will be 4 multiplied by itself 20 times, which is a big number. About a trillion, in fact.

For the dinosaur DNA problem, this is absolutely the wrong thing to try. If you threw a random 20-letter primer into one of your test tubes, the chances are very slim indeed that you'd pick up anything at all. Let's imagine that the complete dinosaur DNA molecule was about a billion bases long. In that case, the chance that the dino DNA would happen to contain any single 20-letter sequence picked at random is roughly a billion divided by a trillion, which works out to be 1 chance in 1,000. You'd have to set up an awful lot of PCR experiments before you found even a single fragment of dinosaur DNA.

The answer might be to use shorter primers. For example, the number of possible DNA sequences with just 6 letters is a mere 4,096 (4 raised to the 6th power). But these primers—instead of not sticking often enough to unknown DNA—would stick too often. If you mixed up a bunch of these 6-letter primers with your unknown DNA, they would be jostling one another to latch onto the DNA at all the places they could. Although it's hard to predict exactly what would happen in such a PCR reaction, these short primers would probably attach to the dino DNA, leaving only short stretches between them—intervals perhaps only 5 to 10 bases long. PCR would then reconstruct just the stretches between primers—which means you would get millions upon millions of extremely short fragments. That would be tough to handle.

A reasonable compromise would be to use random primers that have a good chance of appearing once or twice in the hypothesized billion-letter specification of dinosaur DNA. It turns out that if you use random primers of 15 letters, there are a little more than a billion different possibilities (1,073,742,824, to be exact), which means that each one would have a good chance of occurring somewhere in your dinosaur DNA. This would tend to give you fragments a few hundred bases long—which is about the maximum you can expect, given what we already know of actual DNA extraction from ancient insects and plants.

But does this mean you'd have to do a billion separate PCR operations? If each PCR operation took a day to set up and perform, the task would keep a million lab technicians occupied for a couple of years. But here's one of the few instances in this enterprise where something that seems daunting turns out to be easier than you might suspect. You can simply throw all the billion-plus 15-base primers into a single PCR test tube with your dinosaur DNA and let the machine run. (You can make the billion 15-base primers in about an hour; a little machine—costing a mere $25,000, if you want to keep track—will do the job for you.) The primers will attach where they will, and PCR will multiply all the fragments of DNA between one primer and the next.

You might be thinking that a billion different primers would be a lot of material to put into one little test tube, but it's not. Molecules are small. Even if you throw in 1,000 copies of each of the billion primers, you've still got "only" a trillion molecules. A sugar cube, by comparison, contains about a trillion trillion molecules of sugar, so your package of a trillion 15-base DNA primers is only something like a trillionth of a sugar cube in size. No big deal.

After PCR has done its work, you should be left with a test tube that contains all your original dinosaur DNA fragments—multiplied many times over. You now have lots of copies of all the dinosaur DNA fragments, but they're still all mixed up together, and you can't go any further until you've isolated each fragment so that you can analyze and identify it.

This is where things get a bit overwhelming. The best way to isolate the DNA fragments would be to use some sort of standard cloning technique, like the one that Wilson and Higuchi used to recover DNA from their scrap of quagga pelt. Each of their DNA fragments was individually incorporated into a bacteriophage, which then multiplied it within a host bacterium. But you can use a simpler and more convenient variation on that theme. Bacteria contain little circles of DNA called plasmids, which perform various housekeeping functions for the bacteria. Geneticists have learned how to incorporate a small fragment of foreign DNA into a plasmid and then to slip the modified plasmid back into a bacterial cell. Once again, E. coli is the bacterium of choice. The bacteria will multiply with the bits of extra DNA inside their plasmids, and it's easy to keep large colonies of the bacteria alive. Now you've got a way of individually storing and multiplying all the bits of dinosaur DNA.

Plasmids can't handle DNA fragments as large as those incorporated into bacteriophages, but for pieces of dinosaur DNA a few hundred bases long, they're ideal. The problems with this technique are logistical. The practical difficulty is that you want to be sure that each DNA fragment gets picked up by at least one bacterium—but at the same time that no bacterium picks up more than one piece of DNA. To accomplish this, you have to adjust the relative concentrations of DNA, plasmids, and bacteria in your mixing and control the amount of time they are allowed to mix. As it was with the bacteriophages, this is something of a hit-or-miss procedure, and that's the main reason for using PCR to multiply the original fragments of dinosaur DNA: if you have lots of copies of all the dino DNA fragments, you can be fairly sure that each fragment will be picked up by a plasmid. With plenty of plasmids, you'll have a better chance of ensuring that at least one copy of each fragment gets lodged in at least one bacterium. Somewhere in your garden of bacteria, there should be at least one colony containing any given fragment of dinosaur DNA.

Still, there's an awful lot of dino DNA bits to be accounted for. If the entire (hypothesized) billion-base genome of a dinosaur were to be broken into segments about 500 bases long, you would have several million fragments—actually, more than that, because each of the original copies of the dinosaur genome in your meat sample would have broken up differently in the extraction process, giving rise to a whole different set of fragments. So let's imagine that you end up with 100 million tiny fragments of DNA, each one tucked inside a bacterium that you're going to grow into a colony.

If each little colony needs a square centimeter of space to

grow in, you're going to need 100 million square centimeters of goo-covered glass. That area, regrettably, is a space 100 meters on a side—comparable to a couple of football fields placed side by side. Of course, you'll have lots of little plates all stacked up—but still, you're going to need a big lab.

When you've grown the colonies, you'll need to scrape them off into individual test tubes for the next step. You can use little plastic tubes a couple of centimeters tall and as thick as your little finger. To store 100 million of these tubes in some kind of automated warehouse (so that each one can be retrieved and replaced by a robotic arm traveling along the rows and up and down the columns), you're going to need a space the size of a large department store.

And think about time and manpower. Every time you have to perform a routine operation on your samples—some task that should take you no longer than scratching your nose—you've got to do it 100 million times. A full year contains only 30 million seconds, so it would take three people a whole year just to do a 1-second operation on every test tube.

Oh, well. At this point, let's just assume that John Hammond is so keen to get his dinosaurs that he's willing to buy huge amounts of real estate, large buildings, and enormous automated machines to shuffle test tubes around, and to employ hundreds, if not thousands, of trained technicians. In fact, it's revealed in *The Lost World* that Site B, the secret island where Hammond's scientists were doing the things we didn't get to hear about in *Jurassic Park*, does indeed contain a lab the size of a couple of football fields. But Site B was mainly devoted to some of the difficulties of getting dinosaur embryos to grow and survive—a stage we're not even close to yet. We need some football-field-size labs just to keep track of all the dino DNA fragments you have extracted.

With all Hammond's resources, you'll somehow get the

job done. At the end of it, there'll be 100 million little test tubes containing, between them, all the remnants of dinosaur DNA that your one sample of amber-preserved dinosaur steak yielded.

The purpose of all this effort, in case it's slipped your mind as you've been trying to keep all those tiny test tubes properly accounted for, is to reconstruct dinosaur DNA. More precisely, you want the complete, properly ordered list of A's, C's, G's, and T's constituting a dinosaur's genetic code. What you've got right now is a huge number of little bits of dinosaur DNA. The first order of business is to find out the sequences of A's, C's, G's, and T's that make up every last one of those fragments. Sequencing segments of DNA used to be a laborious, difficult, and time-consuming procedure, but like so much else in a modern molecular biology lab it has become largely automated. Nowadays, you can take a small amount of water containing your purified DNA sample and drop it into a sequencing machine. Some hours later, the machine prints out a series of A's, C's, T's, and G's. There's your sequence!

How the sequencer performs this trick involves a lot of intricate chemistry that we will spare you here. In essence, the sequencer systematically divides the original piece of DNA into a lot of fragmentary copies, whose ends are tagged with dye molecules of different colors which attach to each of the 4 bases, A, C, T, and G. By separating these fragments according to their lengths and noting the color of their tagged ends, the machine can reconstruct the full sequence of the original fragment. In a matter of hours, current sequencing machines can sequence a piece of DNA about 1,000 bases long. Longer pieces are more troublesome, but here, at last, we get a lucky break.

Since our dinosaur DNA fragments are highly unlikely to be more than a few hundred bases long, each one can be sequenced directly without any trouble.

But you can't dispense with human intervention. Sometimes in the printout from a sequencer you'll see the letter N among the bases, indicating that for one reason or another a base (an A, C, T, or G) in the sequence couldn't be identified. One way around this sort of problem is to do each sequence twice, starting at each end of the fragment. That way you'll get two sequences, one of which should be the exact reverse of the other. With luck, any N's in one sequence will be identified in the other, and vice versa.

A good sequencing machine (costing around $125,000, so John Hammond better have his checkbook ready again) can sequence as many as 64 strands of DNA at once. Let's see: you've got 100 million test tubes, and you can do 64 sequences at once. That means a little under 2 million sequencer runs. Each run takes an afternoon (allowing mornings for clean-up and preparation), so you'll need about 2 million "sequencer days"—that's 2,000 sequencers (costing a total of $250 million) operating continuously for about 3 years. Not to mention all the technical staff you'll need: lab space, chemical supplies, clean-up crews, electricity bills. Even so, the task seems tedious but not infinite. Expensive but not, perhaps, prohibitively so. We're going to have to assume not only that John Hammond is as generous as he needs to be but also that he's a good deal more patient than he appears to be. If he really wants dinosaur DNA, you can supply it. But it takes an awful lot of time and money.

CHAPTER FOUR

THE HUMPTY-DUMPTY PROBLEM

So by now you've isolated and sequenced a whole stack of little fragments of what you think is dinosaur DNA, and you want to put them together to make the complete DNA of a dinosaur—its genome. Trouble is, you don't know what the genome looks like in the first place, so it's going to be difficult to figure out how to put these pieces together in the right order. Moreover, at this stage of the game you can't even be absolutely sure you're dealing with DNA from a dinosaur. If you were lucky enough to find a piece of amber-preserved flesh with some skin attached, you might have been able to determine that it was dinosaur skin by comparing it with fossil imprints of dinosaur skin in various museums. More likely, after you've catalogued all the DNA bits and pieces you might conclude that they don't contain any of the standard sequences you'd expect to find in a mammal or a snake or a crocodile, for example. Ultimately, though, you may just have to plunge ahead without being certain that you've got what you wanted.

In a multibillion-dollar enterprise, more or less the opposite of the one we're faced with, known as the Human Genome Project, scientists in hundreds of labs around the world are embarked on an effort to figure out the complete sequence of the genome of a human being. This is basically a straightforward but enormously time-consuming and costly task, and there has been a certain amount of political and scientific wrangling over whether it's really a good idea. To begin with, as we noted earlier, there's no single human genome that represents the complete DNA of every person on earth; rather, there are about 5 billion human genomes, one for every person on earth. For that reason, the Human Genome Project will in fact sequence the genomes of several humans—a more or less random collection of people, because the labs involved are engaged in sequencing different regions of the 23 human chromosomes, and each lab usually obtains its DNA samples from patients and volunteers in various genetic screening projects. That way, scientists will learn not just the basic sequence information common to all human genomes but also the important regions of variation from one person to another. From a technical point of view, the main difficulty is simply that modern sequencing technology can handle only 1,000 or so bases at a time, and although that number is growing, it's unlikely to grow much beyond 10,000 or 20,000 bases in the foreseeable future. But if the human genome is something like 3 billion bases long, then you're looking at cutting it into a million or so small pieces, all of which have to be sequenced individually and then connected up again. The logistics of this task make the Human Genome Project as much of an exercise in bookkeeping and management as in science itself.

Fortunately, some of the lessons learned in the Human Genome Project can be applied to your Dinosaur Genome Project, in which DNA fragments no more than a few hundred bases long must, like Humpty-Dumpty, be put back together again.

At this point you can take a break from the DNA extraction and sequencing lab, and all those tedious and messy test tube procedures. To reassemble the dinosaur genome, what you mainly need is big computers. Fortunately, John Hammond supplied Isla Nublar with three of the biggest supercomputers his money could buy. With the sequences of all the

DNA fragments safely stored in computer files, you can compare, sift, and manipulate them to your heart's content. The digitized bits of DNA sequences can't be contaminated or lost (assuming there's no Dennis Nedry in your lab trying to sabotage the whole enterprise), and you can play around with them as easily as you can play around with sets of numbers or text files or spreadsheets on a home PC.

The problem of trying to put together a myriad of pieces of dinosaur DNA in the right order to make a correct dinosaur genome is one that Henry Wu is well aware of. In *Jurassic Park* the novel, he shows his visitors a seemingly random string of A's, C's, G's, and T's displayed on a computer screen and declares, "Here you see the actual structure of a small fragment of dinosaur DNA." It certainly looks like DNA—starting off GCGTTGGCTGG and carrying on in that fashion for about 1,400 bases. (In the movie, the cartoon in the visitors' center shows something similar, flashing screen after screen of DNA too rapidly to be read. It's interesting, but a little puzzling, to note that when you slow-mo the video of this scene so you can inspect the DNA display at your leisure, you see that it's not a genuine long stretch of DNA but the same sequence used in the book, repeated over and over.)

As he's displaying the DNA sequence, Wu points out that there's a glitch in it—a little section of 5 bases or so that appears to be missing or unidentifiable.

Now that raises a question. What Wu has put up on the computer screen is a long sequence of DNA, followed by a small break or gap, followed by another sequence of DNA. He implies that somehow he knows that these two sequences match up with just a few bases missing between them—but how does he know that? Conceivably, he may have been trying to sequence a long stretch of DNA, and the sequencing machine figured it all out except for a little bit in the middle where, for some reason, the DNA was unreadable. Such things can happen, especially if you've got a piece of DNA that's not in chemically perfect shape. The sequence Wu is displaying on the computer is almost 1,400 bases long, and it seems to be a stretch that he's picked out from the middle of an even longer sequence.

If that was the case, then Wu might indeed be able to guess that only a few bases were missing. However, this assumes that you've been able to find long fragments of intact DNA, and we've come to the conclusion that this isn't likely at all. It's hard to imagine that such a long piece of DNA would have been preserved intact.

The problem you face in real life is that you have a great many sequences, all a few hundred bases long (at most). Just by looking at two such sections, picked at random out of the whole stew, you have no way of telling whether they came from somewhere close together on the original intact molecule or were millions of bases apart. If you don't know the dinosaur genome already, you can't tell whether two fragments of dinosaur DNA belong close together or far apart—nor can you tell what sort of sequence is needed to fill the space between them.

Which still leaves us trying to understand how to put DNA fragments together correctly. Suppose someone took a single copy of the dinosaur genome, chopped it up into a few million pieces, each a few hundred bases long, and gave you those fragments in no particular order. If you had the sequences of the individual pieces but nothing else—and if you knew nothing about the original genome—could you reassemble the fragments in the correct order? The answer is a straightforward No. The fragments have no markings to indicate where on the original genome they came from or what piece they should be connected to. The universal nature of

DNA is such that any fragment can be joined to any other fragment. (There's one useful detail about DNA that we should mention here. If the bases are beads on a chain, then they're like beads that you link together by clipping a hook on one into a loop in the next. DNA bases have chemical "hooks" and "loops," so that they fit together only one way around. Because of this, you know the orientation of each DNA fragment, and a little piece that reads AATCG is distinct from the fragment GCTAA, which has the same bases in the reverse order.)

A set of DNA fragments from one genome won't do, then, but that's not the end of the story. Realistically, when you find a way of retrieving DNA from amber, what you're going to get is not fragments of DNA from a single copy of the genome—a genome from just one cell—but a mixture of fragments from several cells, each of which held its own copy of the complete genome. This makes life a lot easier, because the fragments will have broken apart in different places, giving you a way of figuring out what pieces go together. To see how that works, let's go back to Henry Wu as he's explaining how he solved the problem of repairing the error in the dinosaur DNA sequence on the computer screen. He uses restriction enzymes, which cut DNA in specific places. With the help of the computer, he picks out a couple of restriction enzymes that will cut the DNA on either side of the break. In effect, he trims back the broken DNA ends to places that have an identifiable sequence, recognized by the restriction enzymes. Then he applies these same restriction enzymes to other dinosaur DNA fragments, chopping them up into smaller pieces. What he's looking for is a short piece that will join to the ends of the broken pieces, patching the gap between them.

Confused? You should be. Wu seems to be mixing up a couple of rather different things here.

Restriction enzymes do indeed slice up DNA into smaller

and more manageable pieces by recognizing particular short DNA sequences. But using restriction enzymes means going back to wet, test tube chemistry. If you had pieces of DNA in solution and wanted to tidy up their loose ends, this is certainly a reasonable thing to do. But when Wu is explaining all this, he's working with computers, not test tubes, and the DNA sequences are now strings of letters stored in the computer's memory—not real, physical molecules floating around in test tubes. At this point, with the DNA fragments sequenced and filed on the computer, it would be foolish to go back to using restriction enzymes to play around with the bits of DNA themselves. If you want to experiment with the fragments, you simply edit the DNA on-screen, as if you were using a word processor. What's more, you would make sure that you kept, somewhere in a file that can't be edited, a true copy of all the DNA sequences you had collected, so that as you played and experimented on-screen you wouldn't lose the original. That's the first rule of working on a computer: make a backup and don't edit the original!

Nevertheless, the trick Wu uses to patch up the DNA is more or less correct—if we forget about restriction enzymes and DNA molecules and think instead about comparing sequences on the computer. Here's what you would really do. First, pick out one fragment from the whole set of fragments and note down the sequence of the 15 bases, say, at one end of your fragment. The number of different 15-letter sequences you can make out of A, C, G, and T is large—just over a billion. That means that the chance of any one 15-letter sequence appearing several times in the entire genome is pretty small.

Now, have the computer look through all the other fragments to see if it can find that same 15-letter sequence in any of them. As long as you're dealing with a sequence that's long enough that it will probably occur only once in the whole genome, you can be fairly certain that when you find the same sequence in another fragment, the two fragments came from the same place on the genome. On the computer screen, you can place those two fragments side by side, so that their shared 15-base segment matches up. That's how you patch together two fragments of DNA to make a slightly longer piece, and you can use the same trick to match the ends of the new piece to some other fragment.

This process of building up contiguous segments of DNA— "contigs," molecular biologists call them for short--is routinely used in the Human Genome Project and in many other genetic studies. In such work, you might use various restriction enzymes to cut up a chromosome in various ways; once you've sequenced the cut-up pieces, you can use the sequence-matching technique to fit the sequences back together again. The only difference in your dinosaur project is that the pieces of dino DNA were cut up not in a test tube by the specific actions of restriction enzymes but by random disintegration over millions of years while the DNA was in its amber tomb.

As long as you have an abundance of fragments, you can sort through them all (or rather get your supercomputer to sort through them all) looking for repetitions of the same long pattern of letters. In this way, you might well be able to put the fragments together in the right order; this seems to be what Wu has in mind, although his explanation of the process is a little odd.

It's painstaking work, to be sure. If the dinosaur genome is a billion or so bases long, and if it's broken up into pieces typically 500 bases long, then you're going to be faced with an awful lot of tiny scraps—tens of millions of fragments from each copy of the original genome. And in order to have a chance of putting the whole genome together, you're going to need 3 or 4 or maybe a dozen original copies, all chopped up in different ways. So your supercomputers will need to search through some 100 million scraps of DNA, each one about 500 bases long. We're talking about maybe 1,000 megabytes worth of information, in the form of all those DNA sequences, plus the computing power you would need to scan and compare them in a reasonable amount of time. A modern home PC can store that amount of data without too much difficulty, but the amount of processing needed to search through every possible matchup between all the DNA fragments is more than your typical PC could manage. A couple of supercomputers should be able to polish the job off pretty quickly, though.

So far, we've made it sound as if you could load all your sequenced DNA fragments into a supercomputer, set the wheels whirring, and, after a day or two, the dinosaur genome would pop out, reconstructed perfectly. But, sadly, this kind of mindless genome reconstruction isn't possible. There are a few complications we haven't mentioned.

Like the human genome, the dinosaur genome was presumably not one long sequence of DNA but was divided up into chromosomes. We have no way of knowing how many chromosomes there should be in dinosaurs when we put their DNA back together again, and, in fact, it's highly unlikely even that all species of dinosaur would have had the same number of chromosomes. Humans have 23 pairs of chromosomes, but gorillas have 21 and chimpanzees have 24, even though gorillas, chimps, and humans are genetically very similar. Among other animals, some kinds of salamanders have hundreds of chromosomes, while the muntjac, a small Asian deer, has but one pair. There may be something meaningful, as far as biology and evolution are concerned, about the number of chromosomes an animal has, but if there is, biologists haven't figured it out yet.

That isn't the whole chromosome story, because we can assume that dinosaurs, like humans—and, indeed, like all mammals, reptiles, and birds—would have possessed two sets of chromosomes, one from each of their parents. When you retrieve DNA fragments from your piece of amber-preserved dinosaur, you will have a lot of little fragments from both these sets of chromosomes. For the most part, that won't make much difference, because the two sets are very nearly identical. On the other hand, there would have been small differences between the two sets, representing the differences between your dinosaur's mother and father. If you have enough fragments to do a convincing reconstruction job, it might be that you would find certain sections of the genome appearing in two very slightly different versions. This would tell you that you had found a place where there was a difference between the dinosaur's two sets of chromosomes, and you would take note of these two versions for later reference. To fully reconstruct your dinosaur, you would have to have both sets of chromosomes in all their intricate detail.

And you might also find some sections that differed quite a bit: in that case, you probably would be looking at the sex-determining chromosomes—the so-called X and Y chromosomes—and then it would be very important to do a good job of keeping the two separate.

The division of the dinosaur genome into chromosomes presents the supercomputer with a new difficulty, because instead of trying to construct one long sequence with two ends, it's looking for an unknown number of sequences with twice that many ends. But there may be a way around this problem. It turns out that chromosomes always end with long, repetitive sequences called telomeres (from the Greek *telos*, for "end"), whose purpose has to do with the need for cells to make copies of the chromosomes when they divide, so that the new cells will have the same genetic code as the mother cell. The biochemical machinery within a cell, just like PCR, needs a "primer" to get it started: telomeres, the sequences at the ends of chromosomes, in effect provide a standard piece of DNA that the cell's copying machinery can latch onto to make copies of the chromosomes. In principle, the new chromosomes should be identical to the old, but as more and more copies are made, there's a tendency for the telomeres to get trimmed down in length a little. Over a long period of time, telomeres can be worn right down to the nub. It's been suggested that one of the things that go wrong as a body ages is that in many cells telomeres are completely trimmed away, so that eventually, when a cell tries to divide and reproduce, a bit of the interior of the chromosome, rather than the disposable telomere, gets lost. This can mean genetic damage to the cell, leading perhaps to malfunction or even cancer.

But that's another story. For the purposes of figuring out dinosaur chromosomes, you would have to tell the computer to look for lengths of repetitive sequences that might be telomeres. Or they might not be. There are lots of places in any genome (human, mouse, fruit fly, and presumably dinosaur) where you can find simple sequences—for example, the 6 letters ATAAGT—repeated over and over and over. These apparently dull sections of the genome, which are known as microsatellites, can be hundreds and even thousands of bases long. They *might* be telomeres, marking the ends of the chromosomes, or they might be stretches from somewhere in the middle of a chromosome.

What's more, if this repetitive sequence goes on for more than a few hundred bases, you can never work out exactly how many repetitions are needed. The reason is that if your dinosaur DNA fragments themselves are never more than a few hundred bases long, you will never find a whole fragment that includes the entire ATAAGTATAAGTATAAGT section complete with whatever distinctive sequences mark its ends.

There are long stretches of any genome that have been given the derisory name "junk DNA" because they seem to serve no obvious purpose. (An example of junk would be the kind of repetitive sequences often used in DNA fingerprinting: the reason they're useful for identification is because they vary a great deal from person to person; that very variability implies that they can't be serving any fundamental biological purpose.) However, not all repetitive sequences are junk. Some long stretches of seemingly redundant DNA may actually be important in determining how the chromosomes fold up inside cells, and that folding, in turn, has implications for the way DNA works in any particular cell. If you don't get the repetitive sequence quite right, your reconstructed chromosome might look as if it had all the genetic information it was supposed to, but the physical arrangement of that information is not what it needs to be for the cell's machinery to read it off correctly.

Another vexing difficulty is that the dinosaur genome, trapped in amber for millions of years, may not fall apart entirely at random. Certain sequences seem to be more fragile than others, and if there were a number of places in the genome that always broke, your reconstruction attempts could never bridge those gaps: it's only by overlapping different sections of the genome, broken at different points, that you can patch the segments together. If some sections of DNA always break apart at the same place, there aren't any overlapping fragments. You will end up with a reconstructed genome still broken into a number of pieces—pieces that could be joined together in a variety of ways.

But apart from all this, the real problem is in assuming that you have all the pieces of a dinosaur genome in the first place. Even though bits of DNA a few hundred bases long can quite plausibly survive, you are going to need millions of such pieces to make a whole genome. Judging by what's been extracted from 30-million-year-old insects, you'll be lucky to retrieve just a few small fragments. The chances that the entire genome is still there, even in tiny pieces, seem remote.

And if there *are* gaps in the recovered dinosaur DNA, you're stuck. If you find two unmatched ends and can't find any fragment that bridges the gap between them, you come to a dead stop. There's no way of knowing whether 5 bases are missing or 5 million; if the whole genome is a billion bases long, even a 5-million-base gap is a small slice of the whole thing. Moreover, if you end up with 4 or 8 or 52 unmatched ends, you're even worse off. You can't tell what goes in the gaps and you can't tell which pairs of ends belong to each other.

Wu's explanation of how he and his colleagues assembled a dinosaur genome, though it has some plausible elements, is ultimately highly implausible. Unless the preservation of DNA pieces is perfect, it just won't work. Even if only 1 percent of the DNA molecule is missing, no dinosaur. One percent is roughly the genetic difference between chimps and humans; it may not sound like a lot, but it makes an enormous difference.

One final point concerning Wu's explanation: In the last decade or so, scientists around the world have begun to identify and sequence so many genetic sequences from so many different organisms that keeping track of them all became a difficult problem. To resolve it, there's now an international storehouse of DNA-sequence data, known as Genbank. Essentially, this is a huge set of computer files, stored at the National Center for Biotechnology and Information, part of the National Institutes of Health. Anyone with a computer and a modem can dial into Genbank. There's a similar DNA-sequence storehouse at the European Molecular Biology Laboratory in Hamburg, Germany, and the two are linked so that you can access either of them with equal ease.

Wu claims that the DNA on his computer screen comes from a dinosaur, in which case you shouldn't be able to find it in Genbank. But a quick search (it took us no more than half an hour) reveals that the DNA Dr. Wu is so proud of is not even remotely exotic. It comes from a still-living and all too common organism, one we've met before: the bacterium *E. coli!*

To be precise, Wu's DNA sequence actually comes from one of the plasmids that's used to store pieces of foreign DNA within *E. coli.* It's easy to see how this plasmid DNA might have been floating around in Wu's lab. What's not so easy to understand is why Dr. Wu would be showing off a piece of bacterial DNA as if it were part of a dinosaur. The plasmid DNA sequence is well known, so it's hardly conceivable that Wu thinks it's really dino DNA. Perhaps Wu is not being quite honest with his visitors. Extracting dinosaur DNA would be a momentous achievement, and after spending all those billions of dollars of John Hammond's money, he wouldn't want to let this valuable information slip into the wrong hands. So he may have put a piece of plain old *E. coli* DNA into the computer for demonstration purposes, figuring (correctly) that none of his visitors would be able to identify it from a quick glimpse on a monitor.

You may have been persuaded by now that putting dinosaur DNA together again, even when you've got all or most of the pieces, is a tricky proposition. But surely we're not going to conclude that Henry Wu is a fraud, and that his claim to have reconstructed dinosaur DNA is a huge fake? After all, the visitors to Jurassic Park are seeing real dinosaurs, which must have come from somewhere. If Wu's method for getting a complete dinosaur genome isn't going to work, and if in fact the DNA he showed to his visitors came from a common bacterium, not an extinct giant, then where did the dinosaurs come from?

We're being a little unfair here. Wu has another trick he uses to repair DNA fragments. As the little cartoon in the visitors' center notes, there were times when Jurassic Park's scientists were faced with gaps in the DNA that they couldn't bridge. In those cases, they resorted to patching up the genome with bits of DNA from another species—a frog, to be precise. This gets the park in trouble later on, when it appears that the frog DNA has caused the all-female population to produce a few male dinos, but we'll get to that later. For the moment, let's think some more about using DNA from one species to repair the DNA of another.

This might sound at first like a poor idea. After all, the reason that one species of animal differs from another is, fundamentally, that they are made from different genomes containing different DNA sequences. On the other hand, the DNA of closely related species is, by the same argument, also closely related. Although a person and a monkey are obviously different animals, they also have a lot in common, not just in their general appearance but also in their internal organs, bone structure, blood chemistry, immune responses, and so on.

To understand this genetic similarity better, we need to know more about how the genetic code embodied in DNA actually works. What is the nature of the instructions contained in an animal's genome, and how are they translated into a working creature? The full story on this is something that science has not yet figured out, but a good deal has become clear since the discovery of DNA's structure almost 50 years ago.

Take blood as an example. The molecule that carries oxygen through your bloodstream from your lungs to your other organs is called hemoglobin. Hemoglobin is a protein, one of several tens of thousands of proteins essential to the structure and functioning of a living body. To be precise, hemoglobin is made of two copies of two different proteins, and these are hooked up in a particular way and centered around a molecular assembly that performs the important task of ferrying oxygen around. The proteins themselves are made of smaller chemical units called amino acids, and the instructions for putting the right amino acids together in the right way are coded into your DNA.

All proteins in your body are strings of amino acids that come in 20 types. For hemoglobin, the instructions for putting the roughly 600 amino acid units together in the right way reside somewhere in your genome, as two individual genes for the two proteins that make up hemoglobin. A gene is a stretch of DNA that carries the code for a string of amino acids.

Now, there are 20 kinds of amino acids, but, as we have already learned, only 4 bases in DNA. So the code can't be the simple arrangement of 1 DNA base for 1 amino acid. It has to be a bit more complicated, and it is, but not a whole lot more. Each amino acid is designated by a triplet of bases. For example, ATG signifies the amino acid methionine and AAA or AAG corresponds to lysine, an amino acid we'll come across later. Any of the triplets TAA, TAG, or TGA serve as something called a stop codon, which marks the end of a gene. It's the sign that you've reached the end of the coded instruction for making one particular protein. Because the number of different triplets you can make from A, C, G, and T is 64 but there are only 20 amino acids to be coded, there's some redundancy in the system. In many cases, the third base of the triplet is meaningless, so that (for example) CCA, CCC, CCG, and CCT all correspond to the amino acid proline.

If a gene is just a stretch of DNA that has the instructions for making one particular protein, then the genome must just be a long list of genes, one after another, right? Well, no. Stretches of DNA that carry the instructions for a sequence of amino acids are called coding DNA. The rest is called, obviously, noncoding DNA. You might be surprised to learn that only about 5 percent of the whole genome (in humans, at any rate) is coding DNA.

So what's the remaining 95 percent for?

The truthful answer is that no one knows the whole story. Some noncoding DNA seems to genuinely deserve the name "junk": it doesn't do anything at all. Other sections of noncoding DNA are purely functional, acting as switches to set off a chain of events that ultimately result in the instructions within the gene being turned into the protein it codes for. Clearly, these regions are important bits of DNA–bits you need to get right. But then there are other large sections of noncoding DNA whose purpose is only now beginning to be understood. Some of the sections are the long repetitive sequences mentioned earlier, such as the 6-base string ATAAGT, repeated over and over. In a broad sense, these pieces of DNA may serve architectural roles–that is, they may interact with other chemical assemblies inside the host cell to hold DNA together in a particular shape. As we'll discover in the next chapter, the way DNA is packaged has a lot to do with the way its genetic instructions are read.

The problem as far as our dinosaur project is concerned is that no one knows exactly how these DNA sections do what they do. Do you need to have a precise sequence, made of a certain shorter unit repeated a specific number of times? Can you substitute a slightly different 6-letter unit? Can you substitute slightly different such units within one long sequence?

There are also pieces of noncoding DNA *within* genes, not just between them. These interpolated sections are called introns, and they may well occupy much more of the sequence in a gene than the coding sections themselves. When a gene is read to make a protein, the introns are in essence passed over, so that only the coding sections end up being translated into amino acids to form the desired protein. But then why are the introns there? They may well be evolutionary remnants: perhaps longer genes arose as smaller genes joined together and some shorter sections were "switched off." Another possibility is that the presence of introns simply slows the protein-making machinery, regulating the rate at which a protein is made. Introns may have more subtle effects, too—in particular having to do with the way DNA is packaged and made physically accessible to the protein-making machinery.

You can see that there's more to reconstructing a gene correctly than just coming up with the right sequence of coding DNA to make the protein you want. Think about hemoglobin again. Every person has two hemoglobin genes, but those genes are not identical from one person to another. Some differences are inconsequential. There's the redundancy in the triplet code, which means that the same amino acid can be coded by as many as 4 different 3-letter codons. Two people with slightly different hemoglobin genes can nevertheless make exactly the same hemoglobin.

But there are also small variations in hemoglobin from person to person. Some of the 600 amino acids in hemoglobin can be substituted for others without any appreciable change in the performance of the hemoglobin itself. Such variations obviously translate into differences in the hemoglobin genes, but these are differences with no real consequence. Then again, some changes are by no means negligible. Sickle-cell anemia, a painful and debilitating disease more common in Africans and those of African heritage than in whites and Asians, is the result of having flawed hemoglobin arising from a single error in one of the hemoglobin genes—a single letter wrong in the coding part. The faulty gene gives rise to severely distorted hemoglobin, resulting in characteristically sickle-shaped red blood cells that do a poor job of carrying oxygen.

There are, of course, also differences in the huge amount of noncoding DNA—differences whose consequences are quite unpredictable. Because true junk gets passed over whenever hemoglobin is made from the hemoglobin genes, the exact sequence of the junk may not matter—except that you still need to worry that an error in junk might cause it to be misinterpreted as not being junk at all, a misinterpretation that could cause problems. Errors in the parts of noncoding DNA that have to do with the storage and interpretation of genetic instructions might also have serious consequences.

This range of possible differences—from serious to inconsequential—makes the job of repairing dinosaur DNA very difficult. If you were trying to reconstruct a human genome by patching it with pieces from a chimp genome, you would probably be able to do a pretty good job. The two genomes are the same, base for base, 99 times out of 100, so filling in gaps in a human genome with chimp DNA would introduce a relatively small number of errors. It turns out, though, that the *coding* parts of human and chimp DNA, which constitute small fractions of the respective genomes, differ by as much as 10 percent; this suggests that you might be able to patch up the junk DNA from one with junk DNA from the other without too many ill effects, but that once you start messing with the genes themselves, you could run into more serious problems.

Henry Wu, on the other hand, is trying to fix up dinosaur DNA with bits of DNA from frogs, hardly a close genetic relative. The choice of frogs is mystifying. If you were trying to do this at all, your best bet would be to look at bird genomes, because birds are thought to be the nearest living relatives of dinosaurs. But even birds aren't all that close to dinosaurs. As species develop, one from another, their genetic codes diverge, becoming gradually more and more distinct as the creatures themselves become physically more distinct from their ancestors. The fact that birds are physically so different from dinosaurs tells you that there must be a lot of genetic differences, too. Nevertheless, there's a common history that can perhaps be exploited; you don't really have any other option but to look at bird genomes.

This is going to be a tall order indeed, but John Hammond won't be very happy if you get as far as reconstructing a large fraction of a dinosaur genome only to tell him that you can't figure out the rest and the project's dead. You have to come up with some sort of proposal, and here's the best we can think of.

You've patched together as well as you can all the fragments of dinosaur DNA you've managed to collect. There's no telling how well this will work, since it all depends on the number and size of the fragments you've been able to recover. The computer may also have provided you with several possible ways of patching some of the fragments together and no indication of which of these is the right way—in which case you have to take into account all possible patching sequences and test them all.

At this point, you go to the birds. That is, you pick a variety of birds, including some primitive ones—the so-called ratites, such as emus and ostriches. You'd also want to include a variety of other birds with varying degrees of closeness to the ratites, for comparison. You then need to sequence completely the genomes of several such birds—as many species as possible. Bear in mind that this would be a task at least as expensive and time-consuming as the Human Genome Project. It will take all your powers of persuasion to get Hammond to underwrite this particular part of the enterprise. He doesn't seem like the kind of person who would be content to wait for a couple of decades while you're sequencing bird genomes.

Oh, and it might be a good idea, while you're at it, to sequence the genomes of some other, less closely related species: lizards or snakes, perhaps; maybe a toad and a turtle.

When you've done all this, you take the dinosaur DNA fragments you've managed to piece together and you compare them with all the bird sequences that you've now obtained. What you're looking for are similar, or perhaps even identical, sequences. If you can find a close cousin to one of your dinosaur sequences in one of the bird genomes, and if it's a sufficiently long sequence to be essentially unique, then you could use that bird genome both as guide to where the corresponding bit of dinosaur gene might belong on the dinosaur genome and as a model for filling in missing pieces in the dino gene.

If you can recognize a single dinosaur gene as it emerges from the reconstruction process, you might want to learn something about its function. You might try, for example, inserting the gene into *E. coli*, to see if the modified bacterium will manufacture the dinosaur protein corresponding to the gene. More elaborately, you might try inserting the gene into the fertilized egg cell of a living animal, to see what (if any) physiological effect it will have on the animal that develops. Because of the interconnected evolutionary history of life on earth, it's not unusual to find similar genes performing similar functions in, say, humans and mice and even fruit flies. Scientists have snipped an important gene out of a developing fruit fly egg, making it unable to grow into a fly, and then substituted a comparable gene taken from a mouse, which allows the egg to develop into a fruit fly properly. Even though the mouse and fruit fly genes are not identical, their structure and function are similar enough to be interchangeable. Scientists have also taken mouse genes known to be involved in eye growth and inserted them at random places in a fruit fly's genome, causing the developing fruit fly to sprout authentic-looking eye structures on its wings or legs or other odd places.

If a foreign gene is taken up into the genome of another species, there's no guarantee it will do anything at all; but if you're lucky, the dinosaur gene will not only become incorporated into the developing mouse or fruit fly embryo but will also have some interesting effect or other. Adding a single dino gene won't cause the resulting mouse to grow scales on its back or develop huge, powerful hind limbs with fearsome claws: those kinds of features are not controlled by single genes but by whole cascades of genes and proteins interacting and feeding back on each other through the complex course of an embryo's growth. But it's conceivable that at least a few dinosaur genes could have some significant effect on the developing host, perhaps as obvious as an alteration in the texture of a mouse's fur or as subtle as a minor change in blood chemistry. Experiments of this sort might well teach you not only about the function of those particular dinosaur genes but also a little bit about dinosaur physiology and biology.

But your principal task, of course, is to create the entire dinosaur genome, and your method has been to identify individual dinosaur genes by placing the dino DNA fragments alongside similar sequences within known bird genomes. The end result, though, will be a sort of fake genome, based on some sort of average or extrapolated version of all the bird genomes, with exact pieces of dinosaur DNA attached wherever you are sure of their placement.

If this sounds like a wildly hit-or-miss procedure, it is. You can only hope that you get enough dinosaur DNA to allow reasonable coverage of the whole genome, because where you don't have dino DNA you're going to have to fill in the gaps with guesswork from the bird DNA sequences. As long as you're willing to make enough guesses, you can come up with a sort of wannabe dinosaur genome. Whether this will correspond to anything like a real dinosaur, or indeed any kind of living creature at all, is something you have to pray for. The less dinosaur DNA you've been able to find, the more your wannabe dinosaur genome is going to look like the genome of a bird with a few dinosaur bits in it rather than a dinosaur with a few bird sections.

If you're going to indulge in extravagant thinking, here's something more farfetched than using bird genomes to help you construct a dinosaur genome. Why not try breeding birds in a highly selective manner, with the eventual aim of creating dinosaurs? That is, you would breed birds so as to bring out, bit by bit, the dinosaur characteristics that still lurk within their genetic constitutions. It would be a bit like breeding dogs, as people have done for centuries, to amplify whatever qualities of speed or size or intelligence the breeder requires.

On their own, dinosaurs bred one generation after another, and as the eons went by, those generations changed imperceptibly into whole new species. In one particular lineage, it is thought, that process led ultimately to what we now recognize as birds. There would have been a series of increasingly birdlike dinosaurs, culminating at some essentially arbitrary point in creatures we would tend to call birds rather than dinosaurs. Occasionally in evolutionary history, fairly pronounced and definitive changes do occur. An interesting example is modern corn, which arose from a grasslike plant with a much smaller ear of seed, like wheat or barley. It took small changes in only a couple of genes to transform this primitive, small-eared corn into something like modern corn, with its large cob. Likewise, some such abrupt but small change in the genome could have replaced primitive downy feathers with much bigger ones—feathers suitable for flight. But that's only one part of what you need to do to change a dinosaur into a bird. Overall, the transformation must have involved a huge number of tiny, barely noticeable changes.

Every step in this long chain of breeding and reproduction consists of two creatures combining their genomes to make offspring. To create dinosaurs from birds we would need to perform this evolutionary sequence in reverse, finding suitable pairs of "offspring" to breed so as to re-create genetically plausible "parents." The idea springs from a suggestion made by the British evolutionary biologist Richard Dawkins in his 1985 book *The Blind Watchmaker*, to the effect that it might be possible to resurrect the dodo by selectively breeding pigeons, say, for increasingly dodolike characteristics.

But even transforming pigeons into dodos—which are, relatively speaking, close cousins—would be a huge task. People have devoted their lives to breeding faster racehorses or redder tomatoes or bigger roses or sleeker greyhounds, but it takes enormous amounts of time and luck to come up with even small changes such as these. If you're breeding the conventional way—by getting two plants or two animals together and letting nature take its course—there's a large element of randomness involved. Two superfast greyhounds might produce only average offspring, because the two sets of parental genes weren't combined in the ideal manner to produce speedy youngsters. To breed dinosaurs from birds this way would be altogether a formidable job.

A direction in evolution—dinosaurs gradually changing into birds—is apparent only in retrospect, and wasn't built into the system at the beginning. The path that presumably led

The extinct dodo and the living pigeon, from Dawkins's thought breeding experiment.

(*Photo by Julius Kirschner/Drawing by Mick Ellison*)

from dinosaurs to birds would have been one possible path, selected by evolutionary pressures, from countless possible paths. To reverse evolution—to go back the other way—you need to retrace that single path accurately, avoiding the myriad sidetracks and dead ends. If you did manage to produce some weird-looking offspring, could you tell whether it was closer to a dinosaur than the birds you created it from, or further away?

And here's one last consideration: As noted, birds are presumed to have descended from one particular branch of the dinosaurs, which gave rise to creatures that began gliding short distances on the feathery extensions of their limbs. The split-off of birds from dinosaurs happened long before the dinosaurs themselves became extinct, so there's a wide range of dinosaur species you can't hope to re-create by breeding birds. Those unattainable dinosaurs left no living legacy. You can try producing dinosaurs corresponding to the ancestors of birds, but not those others. And you want to know which are the particular dinosaurs you might produce by breeding birds backward? Well, wouldn't you know it—the nearest cousins of birds are thought to be the raptors, the villains of Jurassic Park!

CHAPTER FIVE

MAKING BABIES

*H*ow much guesswork did Henry Wu actually resort to? There's a little clue in *The Lost World*. The mathematician Ian Malcolm and his new colleagues—an arrogant Berkeley paleontologist named Richard Levine, and Sarah Harding, a zoologist specializing in the behavior of animal herds in Africa—are exploring Site B, the abandoned secret dinosaur hatchery on Isla Sorna, another island off the Costa Rican coast. This is "Hammond's dirty little secret," Malcolm explains to his companions—the place where all the experimentation and trial and error and mistakes went on, behind the scenes. The account they were given by Wu back on Isla Nublar about going easily from DNA to dinosaurs was "too good to be true."

Malcolm comes across a series of memos written by the first scientists to grow dinosaurs successfully. There is a sequence of letters in one of the memos; it starts MEFVALGGP and goes on for several lines in the same vein. These letters stand for amino acids, so this sequence represents a protein, not a segment of DNA, and in the memo the protein is called an "erythroid transcription factor"—eryf1. This is indeed a real protein. Lots of creatures have erythroid transcription factors, and eryf1 would be one version of it. The puzzle is that the memo notes that this eryf1 protein came from a dinosaur known as *Gallimimus bullatus*—a real dinosaur—but a quick search through Genbank reveals that the amino acid sequence in the book is identical to the eryf1 protein of a chicken.

Even assuming that modern birds, including chickens, are distant descendants of the dinosaurs, it's unlikely that this one protein would have remained exactly the same through 65 million years or more of evolutionary change; even different species of birds have different eryf sequences. It's very hard to believe that the dinosaur in question just happened to have precisely the same protein as a modern chicken. What's more likely is that this represents an occasion when Wu had to guess. Evidently, the *Gallimimus bullatus* genome had some gaps in it and one of the things missing was a gene for an erythroid transcription factor. For whatever reason, Wu decided that eryf1 from a chicken would fill the bill. Maybe he had tried others and found that they didn't work, or maybe the memo that Malcolm found represents one of the failed experiments.

We did one more piece of decoding. Malcolm also finds a memo listing a string of A's, C's, G's, and T's: obviously a DNA sequence. It was a simple matter to see whether the sequence contained the correct genetic code for the chicken eryf1. And indeed it did—with a little trick. Translating the DNA code, you get 14 more amino acids than are contained in the usual eryf1 sequence. These amino acids are grouped into 4 blocks within the sequence, and the initial letters of their names spell out MARK WAS HERE NIH. Any biologist will recognize the last 3 letters. NIH stands for the National Institutes of Health, which not only conducts a lot of research but also distributes grant money to thousands of scientists across the country. You can't be a molecular biologist and not know what NIH is. But who was Mark, and how did he get into *The Lost World*? Mark Boguski is an NIH scientist who works at the National Center for Biotechnology and Information, and specializes in database storage and retrieval software. He's the person who first identified the aforementioned DNA sequence in *Jurassic Park* as that of an *E. coli* plasmid instead of a dinosaur, and he wrote to

Michael Crichton about the error. Crichton then asked Boguski to provide a more authentic sequence for *The Lost World*. Hence, apparently, the appearance of chicken eryf1. But Boguski added his own signature to the brew.

Malcolm's tour of Site B persuades him that back on Isla Nublar the all-important step of turning dinosaur DNA into real live dinosaurs was accomplished by sleight of hand. Henry Wu, having shown off his molecular biology labs with great pride and explained in some detail how they found DNA and put the dinosaur genome together, whisked his awestruck visitors into an adjacent room where, lo and behold, baby dinosaurs were poking their cute little snoots out of eggshells. And before you'd had time to wonder how all this came about, a baby raptor was nuzzling up to Tim, John Hammond's grandson, and squeaking as appealingly as a kitten. Miraculous!

Here on Site B, the reality was very different. There were many failures for every success—1,000 failed dinosaur embryos for every animal successfully hatched. And even after a successful emergence from the egg, there was a staggering rate of disease and infant mortality. So now, you think, the laboratories of Site B, remaining much as they were when they were abandoned 5 years before, will finally reveal the biggest secret of Hammond's enterprise: How do you turn DNA into a dinosaur?

Unfortunately, you still don't get much of an answer. And for a good reason. Although DNA is indeed the complete genetic blueprint for the animal it belongs to, it doesn't turn into that animal of its own accord, any more than an architect's blueprint will spontaneously turn into a house. Along with the instructions for building an animal—the DNA—you need the right set of tools and materials.

Remarkably, there is one genuine case of ancient life being resurrected from amber. In 1995, California researchers Raul Cano and Monica Borucki announced that they had extracted

JURASSIC MARK

This Lost World eryf1 Sequence:

```
GAATTCCGGA AGCGAGCAAG AGATAAGTCC TGGCATCAGA TACAGTTGGA GATAAGGACG   60
GACGTGTGGC AGCTCCCGCA GAGGATTCAC ATGATAATGA TAACCTCGGA GGATTACGCC  120
ATGGAGTTCG TGGCGCTGGG GGGGCGGATC GCGGGCTCCC CCACTCCGTT CCCTGATGAA  180
GCCGGAGCCT TCCTGGGGCT GGGGGGGGGC GAGAGGACGG AGGCGGGGGG GCTGCTGGCC  240
ACCCCCCAGT GGGTGCCGCC CGCCACCCAA ATGGAGCCCC CCCACTACCT GGAGCTGCTG  360
CAACCCCCCC GGGGCAGCCC CCCCCATCCC TCCTCCGGGC CCCTACTGCC ACTCAGCAGC  420
GGGCCCCCAC CCTGCGAGGC CCGTGAGTGC GTCATGGCCA GGAAGAACTG CGGAGCGACG  480
GCAACGCCGC TGTGGCGCCG GGACGGCACC GGGCATTACC TGTGCAACTG GGCCTCAGCC  540
CTGCTGGTGA GTAAGCGCGC AGGCACAGTG TGCAGCCACG AGCGTGAAAA CTGCCAGACA  660
TCCACCACCA CTCTGTGGCG TCGCAGCCCC ATGGGGGACC CCGTCTGCAA CAACATTCAC  720
GCCTGCGGCC TCTACTACAA ACTGCACCAA GTGAACCGCC CCCTCACGAT GCGCAAAGAC  780
GGAATCCAAA CCCGAAACCG CAAAGTTTCC TCCAAGGGTA AAAAGCGGCG CCCCCCGGGG  840
GGGGGAAACC CCTCCGCCAC CGCGGGAGGG GGCGCTCCTA TGGGGGGAGG GGGGGACCCC  900
TCTATGCCCC CCCCGCCGCC CCCCCCGGCC GCCGCCCCCC CTCAAAGCGA CGCTCTGTAC  960
GCTCTCGGCC CCGTGGTCCT TTCGGGCCAT TTTCTGCCCT TTGGAAACTC CGGAGGGTTT 1020
TTTGGGGGGG GGGCGGGGGG TTACACGGCC CCCCCGGGGC TGAGCCCGCA GATTTAAATA 1080
ATAACTCTGA CGTGGGCAAG TGGGCCTTGC TGAGAAGACA GTGTAACATA ATAATTTGCA 1140
CCTCGGCAAT TGCAGAGGGT CGATCTCCAC TTTGGACACA ACAGGGCTAC TCGGTAGGAC 1200
CAGATAAGCA CTTTGCTCCC TGGACTGAAA AAGAAAGGAT TTATCTGTTT GCTTCTTGCT 1260
GACAAATCCC TGTGAAAGGT AAAAGTCGGA CACAGCAATC GATTATTTCT CGCCTGTGTG 1320
AAATTACTGT GAATATTGTA AATATATATA TATATATATA TATATCTGTA TAGAACAGCC 1380
TCGGAGGCGG CATGGACCCA GCGTAGATCA TGCTGGATTT GTACTGCCGG AATTC       1440
```

Translates to This 304 Amino Acid Long Protein:

```
MEFVALGGPDAGSPTPFPDEAGAFLGLGGGERTEAGGLLASYPPSGRVSLVPWADTGTLGTPQWVPPATQMEPPH
YLELLQPPRGSPPHPSSGPLLPLSSGPPPCEARECVMARKNCGATATPLWRRDGTGHYLCNWASACGLYHRLNGQ
NRPLIRPKKRLLVSKRAGTVCSHERENCQTSTTTLWRRSPMGDPVCNNIHACGLYYKLHQVNRPLTMRKDGIQTR
NRKVSSKGKKRRPPGGGNPSATAGGGAPMGGGGDPSMPPPPPPPAAAPPQSDALYALGPVVLSGHFLPFGNSGGF
FGGGAGGYTAPPGLSPQI
```

And Aligns Like This to the 289 Amino Acid Long Chicken eryf1 Protein
Sequence from GENBANK

eryf1   `MEFVALGGPDAGSPTPFPDEAGAFLGLGGGERTEAGGLLASYPPSGRVSLVPWADTGTLGTPQWVPPATQMEPPH`
lost world . . . .

eryf1   `YLELLQPPRGSPPHPSSGPLLPLSSGPPPCEARECV----NCGATATPLWRRDGTGHYLCN---ACGLYHRLNGQ`
lost world . . . . .MARK. . . . . .WAS . . . .

eryf1   `NRPLIRPKKRLLVSKRAGTVCS----NCQTSTTTLWRRSPMGDPVCN---ACGLYYKLHQVNRPLTMRKDGIQTR`
lost world . . . . .HERE. . . . . .NIH. . . . .

eryf1   `NRKVSSKGKKRRPPGGGNPSATAGGGAPMGGGGDPSMPPPPPPPAAAPPQSDALYALGPVVLSGHFLPFGNSGGF`
lost world . . . .

eryf1   `FGGGAGGYTAPPGLSPQI`
lost world . . . .

The dots in the "lost world" sequence mean that it is the same as in the chicken
eryf1 sequence. The dashes in the eryf1 sequence mean that they are missing in
the eryf1 sequence.

The extra amino acids spell: MARK WAS HERE NIH

This figure details how one can uncover Dr. Boguski's "Kilroy
Was Here" trick in *The Lost World*.

bacterial spores from a bee trapped in Dominican amber and had brought these bacteria back to life. A bacterium is a whole lot simpler than a dinosaur, but this surely sounds like an encouraging development for the Jurassic Park project, does it not?

Unfortunately, Cano and Borucki's achievement, impressive though it was, has very little do with regenerating life from DNA, ancient or otherwise. A bacterial spore is a dried out, shriveled up, generally inert object that a bacterium turns into when times are hard. Many kinds of bacteria generate spores that can survive in ice, in the parched conditions of a desert, in caked mud at the bottom of a seasonal lake, and in other inhospitable environments. Frequently, spores are produced in response to drought: when the water returns, the spore springs back to life, reverting to the bacterium it once was. A spore contains the bacterium's DNA, of course, but it includes a whole lot else besides. It represents a kind of suspended animation, and all you need to do to revive it is add water and perhaps a bit of suitable nutrient. What Cano and Borucki did is certainly remarkable, but in essence it's also just a case of bacterial spores doing what they're supposed to do—that is, surviving a drought. Their survival testifies to the robustness of spores and not, sad to say, to any scientific breakthrough in the resurrection of ancient life.

You, by contrast, have dinosaur DNA and nothing else. Or, to be more precise, let's assume that you have stored in your computer a list of all the bases—the sequence of A's, C's, G's, and T's—that make up a complete dinosaur genome, with the whole thing divided up into the correct number of chromosomes, whatever that might turn out to be.

It's certainly possible to turn a genome such as this into a whole animal because that's basically what a chicken does with an egg. A fertilized egg is loaded with the DNA for a new chicken and, as the egg develops, the list of instructions that the DNA represents is transformed from a recipe into a new chicken. The problem is that no one really knows how this trick is accomplished. The egg contains a lot more than just DNA: it's a complex molecular assembly that can read the genetic instructions and make a chicken. Just as you need a team of skilled workers such as bricklayers, carpenters, roofers, electricians, and plumbers, under the supervision of a construction manager or a foreman, to turn a blueprint into a house, so turning DNA into a live creature needs the skills of a vast set of intricate biological workers—enzymes and proteins—working with the proper timing and sequence. In nature, the biological construction crew comes packaged in the form of an egg. So your job now is to reproduce a dinosaur's egg—or at least provide a simulacrum close enough so that your hard-won dinosaur DNA can be set to work.

When you think of an egg, the first thing that comes to your mind is probably one of those oval, hard-shelled things with a runny yolk and white inside that you have for breakfast. To a biologist, the word "egg" has a more specific meaning: it's a single cell, produced by a female, that needs to be fertilized by a male sperm in order to grow into a new creature. Once fertilized, the egg cell starts dividing into a little cluster of cells, and soon that apparently featureless blob of cells turns into an embryo, which grows into an infant animal.

If dinosaurs had been mammals, you would be faced with an insurmountable problem: you would have to get a dinosaur embryo to grow inside the womb of some other similar animal—and no such thing exists (think how big some baby dinosaurs must have been). But dinosaurs were egg layers, and that makes the problem a little easier. Enormously difficult, impossible by present-day standards, but not entirely unimaginable.

The fact that dinosaurs laid eggs isn't speculation: fossil

hunters have found a variety of dinosaur nest sites containing unbroken eggs, egg fragments, and infant and juvenile dinosaurs. One fossil egg on exhibit at the American Museum of Natural History is an odd, conical thing, looking something like an ice-cream cone about 9 inches long and 4 inches across at the big end. Others are more conventionally egg-shaped: the biggest of these is about 18 inches long and almost cylindrical (its species is unknown). Many fossil dinosaur eggs are cantaloupe-size—not too unlike a modern ostrich egg. An egg can't be much bigger than that; if it were, the eggshell would need to be so tough—to keep the contents intact—that a baby dinosaur wouldn't have the strength to break out.

But you can't just throw DNA from one creature into the egg of another and expect something to start growing. Different eggs know different things, and to make a dinosaur from your dinosaur genome, you need to have an egg that's set up to do the things a dinosaur egg needs to do.

Think about what happens as an egg cell develops into an embryo. At first, you have just a single, virtually featureless cell. Once an egg cell is fertilized by a sperm, however, things start to happen: it divides into 2 then 4 then 8 cells, in a cluster. Were this simple pattern of growth to continue, all you would get would be a bigger and bigger blob of identical cells. But if a real creature is to develop, different cells have to start growing in different ways. Some become skin or muscle, others become nerves or liver or blood. The huge variety of cells in a body all grow from the single egg cell—and just how that happens no one really knows.

We do know that the instructions for making all those cell types reside ultimately in the DNA. It's therefore clear that the egg cell has to come equipped with complex mechanisms that can read and translate those instructions correctly. But the process is so intricate and convoluted that tracking down in

Fossil dinosaur eggs and living bird eggs (*left to right*: therapod, saurapod, chicken, ostrich).

(*Drawing by Mick Ellison*)

detail how each step is taken in the progress from egg cell to embryo to independent organism has so far eluded the best efforts of biologists. To grow dinosaurs from the dinosaur genome, you're going to need yet more guesswork and a lot of luck.

To begin, how would you install your dinosaur genome in any kind of cell, let alone a cell that might grow into a new dinosaur? Bacterial DNA floats about in the bacterial cell's interior, in a little glob that includes various proteins and enzymes; these enzymes translate the instructions from DNA so as to make new proteins to keep the bacterium functioning correctly. But bacteria are one-celled creatures, unusually simple. In the cells of anything more complicated, DNA resides in a central structure called the nucleus. If you think of a cell as a bag of fluid, then the nucleus is a bag within a bag. Enclosed within a porous membrane that lets important molecules come and go, the nucleus is where protein-making instructions from DNA are read off, by biochemical machinery that manufactures a molecule called messenger RNA. This RNA—ribonucleic acid, a chemical cousin to DNA—floats out of the nucleus and is picked up by another set of biochemical machinery, which uses the RNA's coded instructions to make the protein corresponding to the stretch of DNA that the RNA was made from.

But even within the cell nucleus, the DNA isn't just floating willy-nilly. It's wound up and stored (to use an evocative image coined by the geneticist and writer Christopher Wills) like great lengths of rope stowed away in the hold of a ship. Coiling it up correctly is a hugely complicated business, which scientists are only starting to learn. If you could unwrap and string together a set of 23 human chromosomes, say, they would be about 3 feet long. And in almost every one of the 100 trillion cells in your body you have a pair of these 3-foot-long genomes wrapped up into tiny bundles far too small to see!

The wrapping begins with a unit called a nucleosome, which consists of a couple of hundred DNA bases wound around an assembly of short proteins known as histones. A chromosome of 100 million bases therefore contains about half a million nucleosomes. How the half million or so nucleosomes wrap themselves up into a 3-dimensional structure remains mostly a mystery. A good analogy is to think of twisting a long rubber band, like one you'd twist to power the propeller on an old toy plane: at first the rubber band just coils around itself, making a tidy spiral structure; but if you keep twisting, the spiral itself starts to coil, giving you an extra level of twisted structure.

If you took one of your own chromosomes and magnified it until it was the width of a very skinny rubber band, it would be about 10 miles long. So imagine two people holding the ends of a 10-mile-long rubber band and twisting and twisting and twisting, until the whole thing is raveled into a ball the size of your fist. And now think about trying to figure out precisely the way it's arranged and how you would describe that configuration to someone else, so that another rubber band could be twisted up in exactly the same way. Complicated, isn't it? At this point, you begin running into some real problems. Although there's a good deal of broad similarity in the packaging of DNA in quite widely differing species, the fine details of chromosome structure are decidedly not all the same. And those details may have something to do with what makes, say, a human a human, and a chimp a chimp.

Moreover, the precise way the genome is stored may have significant effects on what DNA does in a particular kind of cell in any one species. Think about two different types of cell in your own body—liver cells and skin cells, for instance, since they're clearly so distinct in their function. The cells in your liver are constantly manufacturing proteins to keep your liver working correctly, and the instructions for making those proteins come from the DNA in the liver cells. Likewise, your skin regenerates after you cut yourself because skin cells can read off the instructions from their DNA to make more skin cells. The liver cells and the skin cells both contain your *complete* genome, the whole DNA. But in liver cells, only those genes relevant to your liver are read, and in skin cells only the genes needed to make other skin cells are read. If the genome is a huge encyclopedia, then in your liver cells it's open only at the pages marked "liver." How do your liver cells "know" that they're liver cells?

It seems likely that chromosome structure—the way DNA is wrapped up into the nucleosomes and then twisted into a big ball of spaghetti—has something to do with whether genes are read or not in different kinds of cells. Your task is not simply to figure out how dinosaur DNA should be packaged inside a cell nucleus but how it should be packaged inside the nucleus of an egg cell—as opposed to (for example) a liver cell or a skin cell. And that's far from the only problem. There also seem to be biochemical changes that affect the way DNA is utilized in different kinds of cells. Remember that as an embryo develops, what begins as a single egg cell must differentiate into a huge variety of cells as the organism acquires the appropriate limbs, tissues, and internal organs. When a cell takes on a specific function—becoming a fully fledged liver cell, for instance—the change is generally irreversible: that is, once it's a liver cell it can't go back to being any other kind of cell. Such transformations are accomplished by a complex system of chemical switching—a gene makes a certain protein that interacts with other genes: sometimes triggering them to produce their own proteins; sometimes inactivating them, so that their proteins can no longer be made. By the time a cell in a growing embryo has taken on its appropriate characteristics, the majority of its genes may have been permanently deactivated by chemical modification of the DNA itself.

If DNA is to function correctly inside a fake dinosaur egg, it must be correctly packaged, in such a state that the biochemical machinery of the egg can activate the necessary steps, in the right order, to get embryo development rolling.

We're forced here to admit a big gap in our knowledge, and we'll just have to suppose that as biologists and geneticists learn about the way an egg works, they'll begin to understand how you might prepare a genome so that it can be correctly installed within the egg's nucleus. For the time being, the way to do this is a mystery.

Before you get to the problem of packaging DNA into reasonably authentic structures, there's another technological problem to solve. You have the whole dinosaur genome listed as a sequence of bases on your computer, but now you need to turn that list into real DNA—actual physical molecules—so that you can experiment with different ways to wrap it up. That's no easy task. You can indeed make single-stranded DNA, not the double helix, with the same machines you used earlier to compose PCR primers. You type in a list of bases, and after a few minutes the machine gives you, in a little test tube, many copies of the DNA sequence you requested. However, the lengthiest DNA strands you can reliably compose in this way are not much more than 100 bases long: to make a single chromosome, you will need a million such pieces.

With enough of the right machines, that's a big though not inconceivable task. But to make a chromosome, you of course need DNA in its double-helix form—two complementary strands side by side—and you need all the pieces joined together in the right order. There's a cute way that you can do this, and it relies on the fact that one strand of DNA in the double helix is complementary to the other. Using the PCR primer machine, you manufacture a series of 100-base units corresponding to a single complete chromosome strand. You will get a test tube containing short lengths of this single-stranded DNA, divided as follows: bases 1 to 100, 101 to 200, 201 to 300, and so on—all present in numerous copies. Now you set the machine to manufacture the complementary strand of DNA, also broken up into short lengths, but here you break the lengths up differently. This second test tube will contain segments divided thus: bases 1 to 50, 51 to 150, 151 to 250, and so on. To make the complete chromosome, you simply mix the contents of the two tubes together and let the DNA do the work. The segments will attach to their complements, and since the segments overlap, you'll end up with the complete chromosome.

Alternatively, as scientists learn more about 3-dimensional chromosome structure, it may be possible to adapt the molecular machinery from existing cells so as to persuade them to make DNA according to a specified recipe and wrap it up as they go. That might, in the end, be a more dependable procedure than trying to manufacture an enormous length of DNA using test tube chemistry and then resorting to additional lab procedures to wrap it properly.

On to the next problem. We said that different eggs know different things, but the key question is, What things does an egg know and how does it know them?

One major source of an egg's "knowledge" is an enormous pool of maternal RNAs. RNA exists in every cell, as an intermediary between genes and the proteins they code for: the DNA code is transcribed into RNA, and the proteins are made in accordance with the instructions in the RNA. But an egg cell comes equipped with its own crucial set of preexisting RNAs, which will be translated into their proteins once the egg starts developing but before any additional instructions have been read off the DNA itself. This first set of proteins is augmented, moreover, by a preexisting set of maternal proteins; their purpose is to initiate the sequence of events by which the egg's DNA will be transformed into a developing embryo. Without them, and without the maternal RNAs, the whole development process can't get started. The egg cell also contains a chemical gradient that distinguishes its top from its bottom. From the outset, one end of the developing egg is destined to become the embryo's head and one its tail, and this polarity is determined by a chemical signal within the egg which is more concentrated at one end than the other.

In principle, the instructions for making these essential maternal ingredients of the egg ought to reside somewhere in the dinosaur's genome. After all, in the normal course of things, a fertilized dinosaur egg that is destined to become a female must contain the instructions for making the egg cells that a female dinosaur will have in her ovaries in order to make the next generation. But here is a true chicken-or-egg paradox: dinosaur DNA contains the recipe for making a dinosaur egg cell, but you can't read that recipe unless you already have an egg cell! To resolve this paradox, you would have to learn how to interpret the dinosaur's genome with such precision that you could work out just what RNAs, proteins, and other molecules the egg needed to correctly read and translate the other instructions in the genome. That's an almost unimaginably complex task. There's no tidy section of the genome where all the instructions for making an egg are tucked up together; rather, an actual egg that grew inside a real dinosaur would have been the end product of a highly specific sequence of actions within the developing embryo and adult dinosaur, guided at each step by the genome but also constantly modified by feedbacks and other influences within the organism itself. Moreover, the development of an adult organism will be influenced by external factors, such as nutrition, disease, or injury.

In other words, to understand how to make a dinosaur egg cell from a dinosaur genome, you would pretty much have to understand how to make the entire dinosaur—which is the problem you're trying to solve in the first place!

And if that's not daunting enough, here's one more little difficulty. In every living cell, including egg cells, there are blobs called mitochondria, which float around in the cell's watery contents and generate energy to keep the cell working. The problem—a little fact we've had to leave aside until now—is that mitochondria have their own complement of DNA, their own minigenome. They need to make certain proteins in order to function correctly, and the mitochondrial DNA (mtDNA, for short) carries the instructions for making this handful of proteins. Cells typically contain about 1,000 individual mitochondria, all essentially the same, and each one has a piece of

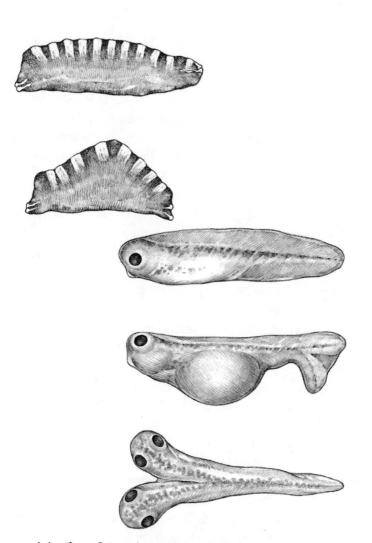

Drosophila (fruit fly) and *Xenopus* embryos that did not get the correct maternal and early embryonic messages. The *Drosophila* embryos are from a mutant line called bicoid and have formed with two butt ends. Needless to say, this embryo dies early in development. Two anomalous early embryonic mess-ups in *Xenopus* are shown: one causes two tails to be formed, and the other causes the formation of two heads. *(Drawing by Mick Ellison)*

DNA in it about 20,000 bases long. That's for your average animal; in humans, mtDNA is exactly 16,569 bases long. At any rate, the total amount of mtDNA in a typical dinosaur cell, we can guess, will be something like 20 million bases' worth (that's 20,000 per mitochondrion, multiplied by 1,000). If the nuclear DNA—the dinosaur genome we've so painstakingly reconstructed—is a billion or so bases long, then it outnumbers or outweighs the mitochondrial DNA by a factor of about 50.

The process of extracting DNA fragments from preserved dinosaur flesh would have also yielded bits of mtDNA along with all the snippets of the true dinosaur genome. Since all the mtDNA has the same sequence, it's reasonable to imagine that these pieces would have stood out prominently in the computer mix-and-match technique you used to reconstruct the genome. So let's just suppose you've succeeded in identifying the sequence of the dinosaur's mitochondrial DNA as well as of its genome. Compared with the genome, the mitochondrial DNA will be short and simple, probably containing no more than 13—at most, 20—individual protein-coding genes. But these genes are nevertheless an important part of what makes a dinosaur work, and if you don't have the right ones, things can go wrong. Following the instructions of the genome, the nucleus of a cell makes a rich variety of proteins, and some of those proteins sail out into the body of the cell and interact with the mitochondria. Because mitochondrial DNA and nuclear DNA have evolved side by side for millions of years, they form a team. A dinosaur cell nucleus generates proteins attuned to interact with the same cell's mitochondria, and vice versa.

This adds yet another complication to the idea of inserting a dinosaur genome into the egg cell of another species. Even though the mitochondria in the egg cell are relatively simple biochemical engines, there's no reason to expect that the non-dinosaur mtDNA in them would interact correctly with the dinosaur proteins emerging from the egg-cell nucleus. It's as though you had put a Buick engine in a VW Beetle and expected all the dashboard controls to work without modification. Incidentally, mitochondria in the egg cell of any creature come directly from the mother. The father makes no contribution. Moreover, the egg-cell mitochondria give rise to all the mitochondria of the growing embryo and, therefore, of the adult animal it develops into, including (if the offspring happens to be a female) that animal's own egg cells.

Another problem: When real dinosaurs mated, the mother provided one set of chromosomes and the father provided another. The two sets would have created a junior dinosaur with genetic elements from both parents but identical to neither. Depending on how successful you were in obtaining dinosaur DNA fragments and piecing them together, you may or may not have a complete specification of two independent sets of chromosomes. At best, you had enough fragments, with enough redundancy, to figure out not only the basic dinosaur genome but also what sort of likely differences existed between the two sets of chromosomes inside the original dinosaur. In that case, you could give your reconstructed dinosaur two independent sets of chromosomes, as a real dinosaur would have had.

But if you weren't able to retrieve enough DNA fragments and you had to fill in the gaps with guesswork and extrapolation from other genomes, you won't have a dinosaur genome you can be confident of. In that case, you will also have no idea how to construct a plausible second genome to "mate" with the first when you attempt to create a new dinosaur.

Why does this matter? You might think that if you had a single set of chromosomes, you could make another set identical to the first and then set them side by side in an egg cell. That might not be utterly natural, since every real creature has two slightly different sets of chromosomes, not duplicates—but never mind nature: would it be workable?

Unfortunately, the answer is probably No. One of the reasons we each possess two sets of chromosomes is that this arrangement guards against a variety of genetic errors. Broadly speaking, you have two copies of every gene so that if one of your chromosomes happens to contain a faulty copy, the other will contain a working one. It's been estimated that everyone has perhaps a dozen potentially fatal genetic errors in his or her genome; the reason they aren't in fact fatal is that there is a good copy of the relevant gene as well as the bad one. But that also means that if you were to take a single set of chromosomes and duplicate it, the faked-up double set may well contain lethal genetic instructions, simply because any flaw in one genome would be faithfully reproduced in the other. And that doesn't even begin to take into account the errors that have crept into your dinosaur genome reconstruction procedure.

This is not good for your budding dinosaur. Nor does there seem to be any way around it, except by trial and error: that is, you have to make small modifications to the genome and hope for the best. But this would be truly a desperate measure! Such random mutations almost always produce a fatal or at least a damaging error, rather than some benign or even neutral outcome, like a change in eye color. Even assuming they had no fatal genetic flaws, your dinosaurs would be clones—all absolutely the same, in every last detail. This might make the herd in your theme park behave rather oddly, since it would lack the usual range of abilities and behaviors found in an animal community.

There's another problem with putting two sets of artificially generated chromosomes into an egg cell. In a normally fertilized egg cell, the two copies of the genome contributed by the egg and the sperm do not act as equal partners. Certain parts of each genome are "silenced," so that the maternal copy takes care of some functions in the developing egg and the paternal copy takes care of others. How this division of labor is arranged, and how precise it must be, remains fairly mysterious. Unless you have learned how to correctly balance the actions of the two chromosomal sets when you load the dinosaur genome into an egg, embryo development may once again fail.

In short, an egg is a hugely complex device, able to read accurately the DNA of its own species but unable to fully interpret any other. Expecting it to do so would be like expecting an English-speaking person to understand French simply because the two languages use the same alphabet: you might recognize some of the words and guess at their meaning, but you can't properly understand an entire book.

You can see that there are several very good reasons why Michael Crichton skated over the matter of transforming DNA into dinosaurs. But let us press on. There are some things you know how to do with egg cells and nuclei. It's even possible to clone whole animals—that is, to grow a new animal that is genetically identical to an original. The big problem is that a technique that works with one species won't necessarily work with another—and, again, no one really knows why.

One of the staples of laboratory experimentation on eggs and fertilization is the creature called *Xenopus laevis*, a South African clawed toad. It's a handy animal, about a foot long at full stretch, and fairly easy to breed in the lab. It owes its scientific prominence to a couple of convenient facts: First, *Xenopus* egg cells are big, a few millimeters across. This makes them easy to see and easy to handle. Second, there was already a minor industry in *Xenopus* breeding before any fancy experiments were dreamed up, because the toad has an odd and unexpected utility: it can be used as a human pregnancy tester! *Xenopus* responds strongly to a hormone made by pregnant women. Exposed to a small sample of a woman's urine, *Xenopus* will release hundreds of eggs if the requisite hormone is there—in which case you know that the woman is pregnant.

So *Xenopus* eggs have long been easy to get hold of. In the 1960s, the British embryologist John Gurdon performed a famous series of experiments that were greeted with widespread disbelief because the results were so unexpected. Gurdon took a bunch of newly fertilized *Xenopus* eggs and inactivated their DNA by exposing them to mild ultraviolet radiation. At that point, the eggs were raring to go but lacked DNA from which to read genetic instructions. Gurdon then inserted into the eggs nuclei that he'd carefully extracted from *Xenopus* tadpole cells. This is a tricky procedure, demanding a careful eye and a steady hand. You can keep an egg cell alive under a microscope, in a dish of water at room temperature, and then use a long, thin, hollow glass needle to inject a new nucleus into it. You push the needle gently into the egg cell the way you can push a soapy knitting needle through the side of a soap bubble without bursting it. With the needle inside, you inject the new nucleus into the center of the egg cell without otherwise disturbing the contents. The process doesn't always work: the egg might burst, or you might damage other parts of the cell. But with some practice, you can get a pretty good success rate.

Gurdon discovered that if he replaced the nucleus of the egg cell with the nucleus of a cell from the gut of a tadpole, an embryo would develop normally and grow into a seemingly normal adult toad. But if he used a nucleus from a cell that was no longer capable of growing (such as a nerve cell, which is unable to regenerate), nothing would happen.

These results were unexpected because scientists had believed that once a cell had taken on its specific role, its genetic machinery was irrevocably altered. It was thought that there was no way of reactivating the whole genome present in such a cell so as to retrieve the complete instruction book for the animal. Gurdon's experiments showed that this wasn't necessarily so; his achievement suggests that the fertilized egg cell somehow has the power to reactivate and translate DNA from some kinds of other cells—which looks like good news for our dinosaur project.

But once again, there's bad news, too. Gurdon's experiments worked just fine with *Xenopus*, but with mammals it's a whole other story. If you try the same procedure with mice—taking a fertilized egg, shutting down its DNA, and replacing the nucleus with one from the gut of a newborn mouse—nothing happens. You can clone toads this way, but you can't clone mice—or human beings, for that matter.

On the other hand, you *can* clone mice if you use a very specific kind of cell called an embryonic stem cell, which must be taken from a mouse embryo at a very early stage—when all the cells are essentially the same and before any distinct tissues and organs have begun to develop. At this point, each of the embryo cells is still able to generate a whole mouse, but as soon as the embryo begins to look even remotely like the beginnings of a real creature rather than just a cluster of cells, that ability is lost.

But a surprise announcement in February 1997 took the idea of cloning mammals to a new level. Scientists in Scotland revealed that they had produced a healthy lamb by inserting the nucleus of an adult sheep cell—taken from a ewe's udder, as it happens—into a newly fertilized egg from another sheep. The cloned lamb is an exact genetic copy of the six-year-old sheep whose udder it was made from—a true clone. Although no one knows quite how the trick works, the key to this success lay in manipulating the udder cells so that they remained stuck at a specific point of the normal cellular life cycle. The egg could then adopt the nucleus from one of the treated cells and start developing into a normal lamb embryo.

No biologist today fully understands why the DNA in some cells from some kinds of organism can be reactivated to generate a whole animal while others cannot. In fact, it's not clear whether the ability lies more with the egg cell or with the introduced nucleus.

There have, incidentally, been occasional reports of human "clonings," one of the earliest being a tale from the 1970s about a certain millionaire who had himself reproduced in younger form so that he could bequeath his wealth to himself. Such tales are outright fiction. But there have also been scientific reports of cloning, in which the story is true but not as exciting as it sounds. In these cases, scientists were able to encourage "twinning" in the laboratory, making a number of fertile and genetically identical eggs from a single original. This is interesting but fundamentally no different from what happens when a woman naturally produces identical twins.

The successful cloning of sheep makes it much more likely that human cloning will one day become a reality. The Scottish researchers—and others, no doubt—intend to see if they can use their methods to make lambs from different kinds of adult sheep cells and to clone other animals, such as pigs or cows. If that works, and even if detailed adjustments have to be made for other species or types of cells, there seems to be no barrier in principle to the possibility that scientists could eventually make an identical copy of you from just a scraping of skin or a tiny snippet of your spleen. We hardly dare think of the ethical difficulties such achievements would bring in their wake.

In *Jurassic Park*, it seems that Henry Wu inserted complete dinosaur DNA into unfertilized egg cells taken from crocodiles, and then inserted the modified cells into fake eggs made of a porous plastic and filled with synthetic yolk and white to provide nutrition for the developing dinosaur embryo. We have to assume that he somehow figured out not only how to package dinosaur DNA into chromosomes and place them correctly in the egg-cell nucleus, but also that he added to the egg cells a full complement of working mitochondria, maternal RNAs, proteins, and so on.

But working with egg-laying creatures, like crocodiles or birds, presents its own set of problems. In such animals, egg cells are held in ovaries and discharged into the reproductive system. A successfully fertilized egg cell travels down the creature's oviduct, dividing and growing. As it moves along, it gets loaded up with the nutritious ingredients that will form the yolk and white of the egg, and it finally passes through a gland that envelops the whole package in a hard, protective shell.

By the time a fertilized egg is laid, therefore, it already contains an embryo too far along in its development for you to be able to turn it into a dinosaur. Its cells have begun to grow and develop, and even if you replaced the DNA in every one of them with dinosaur DNA, too much growth has already taken place for the embryo's identity to be transformed into that of an entirely new animal. You may be able to persuade a crocodile to lay unfertilized eggs, just as chickens are made to lay unfertilized eggs for the supermarket, but that doesn't get you anywhere either. An unfertilized egg lacks the biochemical structures into which you could insert the dinosaur DNA.

Wu and his team presumably began by extracting single

unfertilized egg cells from crocodiles, much as physicians performing human in vitro fertilization must begin by extracting unfertilized egg cells from a woman's ovaries. In human IVF, the next step is to fertilize the eggs with human sperm and then reinsert them into a suitable womb (which may or may not belong to the woman who provided the eggs, depending on medical circumstances).

You and Henry Wu have a more difficult task. You have to remove the crocodile DNA from the crocodile egg cell (or at least inactivate it), replace it with dinosaur DNA instead, and then persuade the egg that it has actually been fertilized, so that it will start to grow. But then what do you do? Wu implies that the yolk of an egg is nothing but nutritious glop, which can be synthesized in the lab to provide an enviroment for the modified egg to grow in. But it's not that simple. If it were just a matter of providing nutrition, you could drop the egg cell into a dish of standard laboratory growth medium and let it rip. That won't work—for essentially the same reason you can't let a fertilized human egg grow in a test tube. Outside a female womb, a human egg is cut off from the physical and chemical support it needs. Likewise, your would-be dinosaur egg cell needs to be properly implanted inside an egg, with a yolk containing a variety of maternally provided ingredients essential to proper embryonic development. It's also unlikely that you could place the modified crocodile egg cell, now containing the dinosaur DNA, into an unfertilized crocodile egg. A developing egg cell must be correctly incorporated into the egg as the egg is manufactured inside the female's body. If it's plopped into an unfertilized egg at some later time, it has no way of connecting up with the egg's contents.

Here's the best strategy we've been able to come up with. First, instead of using crocodiles as your egg source, you would use ostriches. Dinosaurs are closer, in evolutionary terms, to ostriches than they are to crocodiles, so dinosaur DNA might conceivably be a little more compatible with ostrich egg cells than it would be with crocodile egg cells. More important, since you will be attempting to grow dinosaurs inside them, ostrich eggs are bigger than crocodile eggs!

Oh, and you'll have to explain to John Hammond that in addition to all the other facilities on the island you'll also need an ostrich farm.

Following Wu's example with the crocodiles, you would begin by extracting unfertilized egg cells from an adult female ostrich and replacing their DNA with dinosaur DNA. But rather than trying to make an artificial egg, you would reinsert a modified cell back into the ostrich's reproductive system, in the hope that the ostrich will accept it, pack it up in an egg with the requisite yolk and white, form a shell around it, and lay it. You may have to give the ostrich some sort of hormonal drugs to kick its egg-making system into action, but in a general sense these are the same sorts of problems physicians face when they perform in vitro fertilization for humans.

The advantage of this procedure is that you get the ostrich itself to do all the mysterious things that need to be done for an egg cell to start turning into an embryo. Once the egg has been laid, you'll take it from the ostrich and into the laboratory, to see whether the experiment has been a success. For now, you just have to cross your fingers.

Getting from DNA to dinosaur is by far the hardest part of the entire *Jurassic Park* problem. But even if no one knows today how to fool an ostrich egg into making a dinosaur, that doesn't mean no one will ever know. Understanding embryo development—understanding, that is, the precise sequence of events by which the egg cell activates certain genes in the genome, and how those genes feed back on each other and activate or switch off other genes—is one of the consuming problems of biology today. We can only assume that one day the process will be better understood. Then, perhaps, you'll be able either to adapt an ostrich egg so that it can grow a dinosaur embryo from dinosaur DNA, or—even more extravagantly—you'll create a synthetic egg designed to match up perfectly with a dinosaur genome. At that point, we can return to *Jurassic Park*, with little dinosaur embryos growing inside their laboratory eggs.

CHAPTER SIX

IT'S A GIRL!

After Henry Wu has explained his lab procedure for reconstructing dinosaur DNA, Alan Grant asks him how he knows what kind of dinosaur it encodes. "We have two procedures," replies Wu. The first is something he calls "phylogenetic analysis." That's basically the idea of trying to figure out a family tree of evolutionary relationships for genomes; if you know enough about the genomes of birds and other modern creatures, and if you know how they all fit together on the evolutionary tree, you can guess where, in this big picture, a dinosaur genome would fit, and what particular dinosaur it would correspond to.

Scientists are beginning to understand such relationships. When Allan Wilson and Russ Higuchi extracted DNA from a quagga pelt, one of their goals was to understand how this extinct animal was related to horses and zebras. The evolutionary links between humans, chimps, gorillas, and other primates have been illuminated by DNA studies. On a molecular scale, scientists have compared different versions of certain related proteins found in a number of modern species in order to take a stab at identifying the ancestral protein from which the modern versions derived.

But these are relatively modest studies. For animals as disparate as extinct dinosaurs and modern birds, you're unlikely to get enough information this way to examine a reconstructed genome and say what dinosaur species it came from. Even in the case of human and chimp genomes, which differ

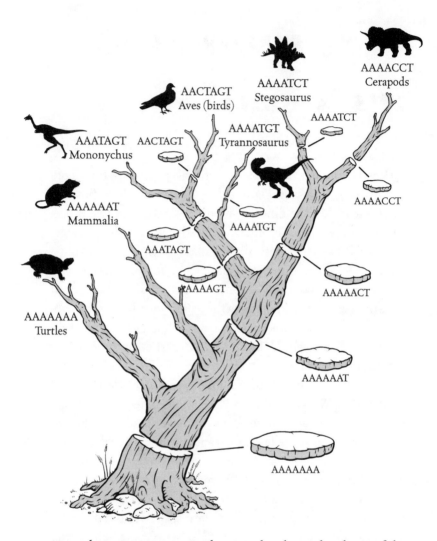

An evolutionary tree, again showing the close relatedness of dinosaurs and birds. This time the tree shows hypothetical dinosaur DNA sequences and how Wu's "phylogenetic mapping" works. Cross sections are cut out of the tree and represent the DNA sequence changes that the common ancestor of all animals above the cross section developed during evolution.

(*Drawing by Edward Heck*)

by no more than about 1 percent, there's no way of understanding why, from the small differences, a chimp looks like a
chimp and a human looks like a human. There are just too
many unknowns in the relationship of genome to animal, and
in the way animals and genomes change over the course of
time. So you'd have to depend on Wu's second procedure,
which is to "grow it and find out what it is." You've got the
DNA, you've figured out how to put it into an ostrich egg and
start it growing, so this is exactly what you do next: grow it and
find out.

At this point, you hope, you can let nature take its course.
You've struggled to assemble a complete dinosaur genome—or
at least what you think is a complete dinosaur genome—from
tiny scraps of ancient DNA; you've fiddled and tinkered with
ostrich eggs to persuade them to take up this alien DNA as if it
were their own; and you've fooled an ostrich into laying an egg
that has a little dinosaur embryo growing inside it. Now all
you have to do is stand back, keep the egg warm, make sure to
turn it over or roll it around regularly, and wait for a little dinosaur to poke out of the ostrich egg's tough shell. Turning the
egg over periodically is important: if you don't, the developing
embryo will adhere to the inside of the shell, causing devastating physical defects. Although you're trying to help the
process along as much as you can, you hope that from this
point on, the intricate biochemistry of the egg itself is doing
all the hard work for you.

So you hope. Whether things actually run quite so smoothly is another matter. In order to see what's happening, you snip
out a window in the eggshell, using fine surgical scissors and
being careful not to break the shell (in which case, you get out
the frying pan and call in your colleagues to share a big
omelette). You cover the window with a piece of transparent
plastic as tough as the shell itself. You illuminate the egg from

behind with a strong light, in hopes of seeing the contents silhouetted, and you watch nervously for the first twitchings of life.

In all likelihood, you'll have taken the new-laid egg from the ostrich, rushed it into the laboratory, and cut open a window to see—nothing at all! For one reason or another, your attempt to implant an egg loaded with dinosaur DNA back into the ostrich has failed. The hormonal treatment you used to persuade the ostrich to accept the implant may instead have stimulated the production of an infertile egg. (You take care to keep your ostriches away from any males, so there's no chance of a fertilized egg being laid.) But even if the implanted egg cell wasn't rejected, it may simply fail to grow, and even if it does grow it may not be packaged into an eggshell. Fertilization done the way nature intended is not itself a surefire procedure. There's a lot about fertilization and embryo growth that scientists don't understand: sometimes it works, sometimes it doesn't, and most of the time you'll never know why.

But don't throw that egg away yet. Even if a dinosaur embryo was beginning to grow, you wouldn't necessarily be able to see it straightaway. When a bird embryo begins to grow, it starts as a disklike layer of cells nestling on the surface of the yolk. Needless to say, you don't know in advance what a primitive dinosaur embryo ought to look like, but it, too, would presumably begin as some sort of structure lying on the yolk. Unless you were lucky enough to put the window into the eggshell just where this little cluster happened to be, there's no guarantee you could see it. It's big enough to see with a moderately powered microscope—but it might be on the other side of the yolk. You can often see the outline of a chicken embryo once it's a few days or a week old, by holding the egg up to a light. But an ostrich eggshell is thicker, and the yolk itself larger and more opaque. Chances are you won't see anything for a while even if the experiment has worked.

On the other hand, if a week or two goes by and you still don't see anything, you will have to conclude that you've got a dud. You must be prepared, as the technicians of Site B were, for much more disappointment than success. But let's be optimistic: after you've peered into 1,000, or 10,000, ostrich eggs, you find one in which something seems to be growing. Let's say, too, that you were lucky enough to put the window in the right place, so that when you look inside you can see a tantalizing cluster of cells on the surface of the yolk. Even then, you're still not in the clear. The task is not only to get the embryo to grow but to get it to grow in accordance with the instructions from a dinosaur genome—something the egg isn't naturally equipped to handle. It's possible that the cell cluster has begun to divide and grow aimlessly, under no genetic control. Probably this sort of growth will just fizzle out after a while, and you'll be left with another dud.

The key event in the development of an egg—the event that tells you that there's a chance a living creature will emerge—is called gastrulation. This milestone, not birth or marriage or getting your driver's license, is what the British embryologist Lewis Wolpert has deemed the most important moment of anyone's life. It's the moment when a seemingly featureless clump of egg cells turns into an object with a perceptible form.

Gastrulation normally occurs when the original egg cell has grown into a cluster of a thousand or several thousand cells, the exact number varying a great deal from species to species. As you look in on the developing egg, you'll see that the clump of cells has taken on a characteristic shape: a flattish disk for birds, a hollow cylinder for mammals, a fat and lopsided ball for *Xenopus*. If you happen to be watching at just the right moment, you'll see the surface of this cluster start to dimple or distort, as if an invisible finger were pressing upon it. The distortion will grow until the cell cluster looks as if it were trying to turn inside out, and its two opposite sides will fold toward each other and join, so that a seam develops. The newly enclosed section is destined to turn into the animal's gut, hence the name "gastrulation" for this process (from the Latin *gastrula* for "stomach").

How soon you expect gastrulation to happen depends on what kind of animal you're dealing with. In flies, it happens a few hours after the cell first divides; in mice and chickens, in about a day; in humans, as many as 4 days may pass before the crucial moment. If your dinosaur is one of the big ones, it might take even longer than that, so you'll have to be patient for several days, keeping your fingers crossed that an egg cell that's begun to divide and reproduce will eventually take this important step.

After gastrulation has occurred, you can see that what was once a fairly simple structure, with nothing more than an inside and an outside, has taken on a more complex appearance. The "outside" cells that are still on the outside will end up turning into skin cells, among other things; the "outside" cells that were folded into the structure will develop into the lining of the gut or will turn into spinal cord structures, depending where exactly they are in relation to the fold. How these changes are directed, and how cells that were originally alike are pushed into various growth patterns and ultimately form quite different cells in the adult creature, is a complicated story that scientists are only beginning to understand. In the case of the nematode *Caenorhabditis elegans*, a (not particularly elegant-looking) worm a millimeter long, this process has been completely mapped out. An adult female *C. elegans* has exactly

1,031 cells. By burning out individual cells with a laser at different stages of embryo development and observing what features of the adult worm thereafter fail to appear, scientists have determined the "lineage" of every cell in the adult body.

In the course of embryonic development, it's also important that some cells be programmed to die at a specific time. Having performed some crucial intermediate step—paving the way for the appearance of an eye or the formation of the stomach, say—one cell or group of cells will die on cue. If they don't, the embryo will retain some sort of transitional structure that no longer has any purpose and may well get in the way of a further stage of development.

The signals that tell one cell to become skin, another to become gut, and still another to die are fundamentally chemical in nature. A protein manufactured by one cell influences the genes in another, steering it to change its identity; that cell, in turn, influences others, and, perhaps, has a feedback effect on some of the cells that caused it to change its form in the first place. Embryonic development essentially depends on genes in one cell making proteins that influence genes in another, and all those influences are tied together in a complex and changing network of interactions and feedbacks. In *C. elegans*, we now know the route all those different cells take, but we don't know all the chemical signals that steer them in one direction or another. Humans, by contrast, have trillions of cells, and mapping out the developmental route by which they all become the right kinds of cells in the right places is a task of daunting complexity.

Given our present state of knowledge, it's certainly impossible, solely by inspecting the sequence of a complete genome, to make any sort of prediction about what kind of organism will develop and how. The answer is buried in there somewhere, but the way it unfolds with the organizational help of the egg cell is quite beyond our present powers of understanding. You simply have to trust that the egg has properly assimilated the dinosaur DNA, and that the egg knows what it's doing. Once gastrulation has taken place, you do know that the first chemical signals to start the developmental process from cell cluster to embryo to organism have taken effect, and that tells you that instructions from the DNA are having an influence. Whether it's the correct influence, with all its proper bells and whistles, remains to be seen.

At this early stage, it's hard to tell one embryo from another. The developing clump of cells is showing signs of a backbone and gut, and you can see the formation of limbs, fore and aft, and a head. But those basic features, when the embryo is just a few days old, don't look very different in a chicken or a human or an elephant. We all have a backbone, a gut, four limbs and a head. At this stage, most embryos also appear to have a tail—in humans, it disappears in due course, except for the bony remnant called the coccyx.

In the nineteenth century, when even less was understood of embryo growth than is understood today, the German biologist Ernst Haeckel came up with a resonant phrase to describe the similarity shown by all newly developing embryos: Ontogeny recapitulates phylogeny, he declared. Phylogeny refers to the evolutionary relationships among species; ontogeny refers to the growth of an individual embryo. Haeckel's idea, put simply, was that when an embryo develops, it goes through a series of stages corresponding to the evolutionary history of the creature. If human embryos first resemble little fishes and then develop limbs with webbed digits before finally turning into creatures with arms and legs, this is because life began in the sea, crawled onto land, and eventually became *Homo sapiens.*

As it turns out, Haeckel was taking appearances a bit too

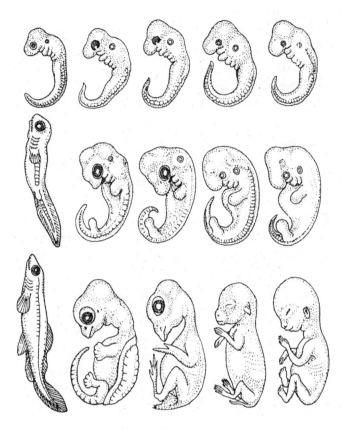

Classical nineteenth-century drawing of the embryos of several vertebrates. Note that early in development all the embryos look pretty much alike; only in later stages do the forms begin to diverge. This pattern is what prompted Ernst Haeckel to claim that "ontogeny recapitulates phylogeny."   (*Redrawn by Rob DeSalle*)

literally. It's not strictly true that embryo development represents a sort of fast-forward summary of evolution. But he was not altogether wrong either. Nature tends to be economical, and getting a featureless egg cell to grow into a complicated, variegated creature is an enormously intricate business. Once DNA "learned," via evolution, how to instruct an egg cell to grow a limb that turned into, let's say, a flipper, that valuable set of instructions was stored securely in the genome. If, over a period of millions of years, a creature with flippers evolved into a creature with legs, it was done through countless tiny modifications of the genetic instructions encapsulated in the genome. The process for making a leg is an elaboration of the earlier process for making a flipper. And because, whether you're making a leg or a flipper, you have to start off in the same general sort of way—by sprouting a little bud from a blob of cells—ontogeny does to some extent recapitulate phylogeny. It may also be that certain developmental tasks, such as the generation of a limb, are so difficult that evolution hits on essentially the same solution independently in many different species.

At any rate, your dinosaur embryo is not going to give itself away at the outset. If you get to the point where an embryo is showing signs of a backbone and gut, of limbs and a head, you can be hopeful that some reasonable set of instructions from the genome is pushing the growing cluster of cells in a particular direction. But you still can't be sure whether you're going to get a dinosaur, or a dinosaur with some ostrich bits in it, or a mixture of different kinds of dinosaur (because there turns out to be DNA from more than one dinosaur in the set of fragments you painstakingly put together), or a dinosaur with bits of chicken and emu and eagle and sparrow (from all the bits of additional DNA you had to use to patch the whole thing together).

Once again, you can only watch, as patiently as possible, and hope. If a healthy embryo is growing and the ostrich egg is working correctly with the imported dinosaur DNA, there should be changes in the appearance of the yolk. You'll be able to see these changes, at this stage, with the naked eye. The yolk will darken and blood vessels will extend into it from the developing embryo. At first, the cells of the embryo absorbed necessary nutrients from the surrounding yolk, but as the little creature begins to grow, it has to be able to pull in and distribute nutrients appropriately to the various kinds of cells it contains. Functions are becoming more precise and specialized, as they must for the embryo to turn into a self-sufficient organism capable of independent life.

The embryo continues to grow and the yolk diminishes as the nutrients it provides are used up. Through the plastic window, you see the growing creature take up more and more space, with its limbs tucked and its head curled beneath its body. As the embryo grows, you begin to anticipate the moment when it's big enough to start breaking out of its shell.

All this time, of course, you've been taking special care of the egg itself. You can't just leave it lying around on a benchtop in your lab, unguarded. The main thing is to keep it warm. In Wu's laboratory, eggs are set out on large trays, in a room that's kept at 99°F and high humidity—conditions meant to mimic the Jurassic climate that prevailed in dinosaur stamping grounds. The oxygen concentration in Wu's lab is a little higher than normal, too—33 percent of the atmosphere, rather than the 21 percent of the present day.

There's some evidence, mainly from the chemical composition of rocks, that oxygen may indeed have been more abundant in the earth's atmosphere in the time of the dinosaurs—but how much more abundant is not known. Recent evidence has come, appropriately enough for our story, from bubbles of air trapped in amber. When carefully extracted, those little pockets of ancient atmosphere were found to have high concentrations of oxygen—up to 30 percent. But whether air trapped in amber is a true sample of the ancient atmosphere is debatable: though amber is impervious to water, it's still somewhat porous. Later experiments indicated that over a period of just millions of years, let alone 150 million, oxygen could seep in and out of pieces of amber, altering the proportion, so that you could never tell how much oxygen was in there in the first place. The notion that the atmosphere in the distant past was significantly richer in oxygen than it is today is not now widely believed. It did, in its short existence, spawn yet another dinosaur-extinction theory: the big animals were used to a superabundance of oxygen, and if forest fires ignited by the giant asteroid impact sucked a lot of oxygen out of the atmosphere, they may have asphyxiated.

But that's by the way. It would perhaps be a good idea to keep the eggs in the hatchery warm and moist, but although eggs do indeed "breathe" oxygen through their shells, it's unlikely they were all that sensitive to the precise amount of it in the air around them.

The one factor that would most likely be of genuine importance here is temperature. Obviously, cold eggs don't develop and hot eggs cook. You want the temperature to be just right. But there might be more to it than that. The egg temperature of some lizard and turtle species controls the sex of the animal that hatches: an egg that's kept cool will develop into a girl; one that's kept a little warmer will be a boy. Henry Wu, as he explains at some length, is anxious to make sure that his dinosaurs all turn out to be females, so that they can't breed. He neuters them with radiation later to make sure they're sterile, but he also controls the development of the egg with hormones.

As Wu points out, "all vertebrate embryos are inherently

female." That might sound a bit odd, since you know that in humans, for example, females have two copies of the X chromosome, while males have one X and one Y. Isn't it the combination of chromosomes that makes a girl a girl and a boy a boy?

Only in part. When a human embryo is first developing, it is in what's called an "indifferent state" as regards sex: you can't tell what it's going to turn into. But at a fairly early stage, male and female hormones (testosterone and estrogen, plus some others) have to switch on to ensure that the appropriate genital organs appear. There's a rare and bizarre genetic anomaly in humans called pseudohermaphroditism, in which youngsters appear to be female, but at puberty begin to sprout male organs. These people have X Y chromosomes, so they ought to be males, but they also have a genetic defect that greatly reduces testosterone production during fetal development. As a result, the embryos don't get the correct hormonal trigger, don't develop male organs, and end up as apparent females by default. Puberty for these people is more of a surprise than it is for the rest of us. As they hit their teens, a new source of testosterone production kicks in, and all the developmental machinery that had been put on hold springs to life. Someone who had seemed to be a girl turns into a boy and then into a man—a fertile and generally normal man. There's a village in the Dominican Republic where this strange course of events was once common and regarded as perfectly routine. It seems that most of the inhabitants of the village were the descendants of a local woman and a wayward Spanish sailor who had this odd affliction. Recently, as the village inhabitants have begun to disperse and mix with people from outside, the incidence of pseudohermaphroditism has declined.

Whether dinosaurs might have suffered from the same problem or other gender confusions altogether, no one can say.

Although the division of the world into males and females is widespread among all species, the way the division is accomplished is remarkably varied. Quite a few species of fish can change sex in both directions, on a number of occasions during their lifetimes. Such fish have the ability to generate male or female organs if prompted by an appropriate hormonal trigger. The grouper, for example, swims in schools consisting of one male, bigger than all the rest, and lots of females. The male emits a hormone that perfuses the water and ensures that the accompanying females stay female. But if for any reason the male disappears, then his controlling hormone disappears too, and the females begin turning into males. The biggest one wins the race, and as soon as her—or his—transformation is complete, he starts generating the same hormone to keep all the smaller fish female. This system guarantees that in any group of groupers, there's always one big guy at the top.

What about birds? They're the closest relatives to dinosaurs, so their method of gender determination may well be the one the dinosaurs used. It turns out that the bird system is a sort of mirror image of the mammalian system: the males have two copies of the same sex-determining chromosome and the females have two different ones. If dinosaurs worked the same way, Henry Wu would have faced an additional problem in making females. If he has only one copy of a single genome, he probably can't tell whether it includes a male or a female sex-determining chromosome. And if he makes up the full genetic complement of a fertilized egg by putting two identical copies of this single genome together, he may or may not end up with a creature that develops properly. He certainly won't end up with a female. If the genome contains the male sex-determinant, then doubling it will give him a correctly endowed (physically as well as genetically) male animal. But if the sex-determining chromosome was for a female, the doubled genome probably won't produce a viable animal of either sex because real females need not just a correct copy of the male and female sex chromosomes but also the various other genes residing on the male chromosome, which the doubled genome hasn't got.

Then again, we can't be sure that dinosaurs would have worked the same way as birds. There's such a wide range of sex-determining factors, and birds are so distantly related to dinosaurs, that we can't expect dinosaurs to conform to any particular pattern existing in modern animals. It's possible that Wu's method of making female dinosaurs by withholding the necessary male hormones from the embryos would work, but even that assumes he can determine the appropriate combination of male and female hormones to use at the appropriate times in embryo development. And Wu's method might demand a greater disruption of embryo development than he would like. If the reconstructed genome works as intended, it should have the instructions for switching on hormone production at the right time. It's not, as Wu suggests, simply a matter of withholding male hormones from the embryo, because the embryo itself ought to be producing the hormones it needs. You'd have to interfere somehow with the developing embryo at the appropriate moment. But how would you know, the first time you grew a dinosaur, what the appropriate moment was?

Still, there's one surefire way of making the dinosaurs sterile. Just wait a while and do what vets do to puppies and kittens. You might want to perform this operation on *T. rex* sooner rather than later.

But if all the animals in Jurassic Park were surgically neutered, you'd lose one of the key points in Crichton's story. Because they've gone out of their way to make all the dinosaurs female, Hammond and Wu are astonished and disbelieving when they find that the creatures are reproducing. The explanation, as Alan Grant finally figures out, is that the dinosaurs with bits of frog DNA in their genomes can change sex: females can become males because of some environmental influence.

Like the groupers, certain frogs and amphibians are able to change sex, and so the idea is that dinosaurs with frog DNA have acquired the same ability. This is hard to believe, for several reasons. First, out of all the DNA of a frog, only some relatively small fraction has anything to do with gender determination. It would be a huge coincidence if the bit that Wu used just happened to be a piece that played a role in the frogs' sex-change habits.

Second, what controls the development of sex and, in some animals, the ability of adults to change sex is not just one gene but a whole complex interaction of many genes. Nor would all the relevant pieces of the frog genome be confined to one place. The essential bits and pieces might come from all over the genome—making it even more unlikely that Wu could have incorporated into his dinosaur the regions of frog DNA governing sex change.

Third, even if, by some astronomical coincidence, Wu had inserted all the right (or wrong, depending on your point of view) pieces of frog DNA into a dinosaur, there's no reason to think that the specific actions of those genes, tuned as they are to a frog's biochemistry, would have anything like the same effect in dinosaurs. The signals sent out by the frog DNA would have to trigger a highly specific response in the dinosaurs, and the dinosaurs, because they're normally creatures of fixed sex, might have no way of responding to those signals—or, if they did respond, it would be in some unpredictable way, causing the animals to be infertile or to develop cancer.

The important point to understand is that the genome of any species has evolved all of a piece, so that the action of one gene has highly specific effects on the actions of many others. Snipping a set of genes from a frog and inserting them in what seems the approximately equivalent place in the genome of a dinosaur will, almost certainly, result in an unworkable genome—unworkable because its constituents have not evolved together and thus presumably cannot act in a coordinated manner. The genes of different species, in effect, don't know how to talk to each other.

You've seen that the embryo was starting to grow and take on a shape, and then you waited. And you waited. And then you waited some more. A chicken takes 22 days to break out of its egg after the egg has been laid. Ostriches and emus take about 3 months. You don't know how long your dinosaur's sojourn in the egg will be, because for one thing you don't know what sort of dinosaur it is. If you have one of the giant dinosaurs—a *T. rex* or an apatosaurus—growing inside an ostrich egg, you figure the thing is bound to bust out sooner or later. But what if the dinosaur embryo is a little procompsognathid, one of the chicken-size scavengers that scuttle about Jurassic Park cleaning up the mess? If it's about the size of a chicken, perhaps it needs a chicken-size egg, too. If you're trying to grow one inside an ostrich egg, it might be ready to start breaking out, but instead of reaching that point when it's squeezed tight inside the egg and can't grow any more, it might find itself ready to be hatched but floating around, adrift, in the middle of an egg three sizes too big for it.

As you're peering through the window in the shell, you may get a sense, from the embryo's appearance or its motions, that it's ready to emerge. And there might come a time when

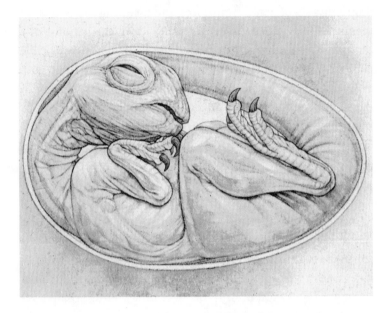

A now-famous dinosaur embryo, found in Mongolia by AMNH
paleontologists. *(Drawing by Mick Ellison)*

you'd want to give it some help, breaking the egg apart and
plucking it out. You wouldn't want to leave a small embryo inside a large egg too long because it could drown. At some
point, the developed creature is programmed to break out of
its egg and start breathing air. If it was unable to break out of
the egg on its own, the switch to air breathing might happen
anyway, and that would be the end of that.

Once you'd identified some of the dinosaur genomes from
the creatures they grew into, you could try performing the in
vitro fertilization trick with birds other than ostriches, so that

the baby dinosaur would be installed in an egg of about the right size. But there would be many failures before you had the first success. According to the memos Malcolm comes across on Site B, Wu's team had 1,000 failures for every successful hatching. Even that seems optimistic, given the magnitude of what you're trying to do here.

But one day—after you've hit upon the proper size egg for the particular genome you're using, and after you've worked out the best temperature for the eggs, you will experience that wonderful moment when the egg begins to crack and a beak (or a snout? or a claw?) begins to poke out.

Now you wait, breathless, as the creature struggles to pop out into the open. Did a baby dinosaur make its exit all by itself, or did its mother have to help? You don't know, and so perhaps, when you see the baby struggling a little, you reach over with a gloved hand to pry apart the eggshell and ease a limb through the opening. At last, the egg breaks fully open, and a little dinosaur—say, about a foot from head to toe—struggles to its feet.

This is the first living dinosaur in 65 million years. It's looking out on a world its parents never saw—a world with racks of test tubes glinting on the horizon, air vents humming, odd-looking white-coated creatures staring at it in wonderment. But the newly hatched dinosaur doesn't know that this world is not the one it was meant to emerge in. It has no expectations, so it takes its surroundings for granted. All it knows is that it's *hungry!* It starts squealing and waits to be fed.

CHAPTER SEVEN

GROWING UP IN A STRANGE WORLD

*B*aby velociraptors seem to be pretty darn cute. In the hatchery, a newly hatched raptor snuggles up to Tim, John Hammond's grandson. Later on in Crichton's novel, when the park is coming apart at the seams, Tim and his sister, Alex, are back in the visitors' center and come across the same little infant, starving and desperate for affection. It clings to the kids pitifully.

But now Tim and Alex are being stalked by full-grown and ferocious raptors, who are not nearly as cuddlesome as the baby. The animators of the movie took some trouble to make the raptors look evil—by human standards. The beasts have dark, shifty eyes and elongated snouts; they move abruptly, then pause menacingly; they make conspiratorial eye contact with each other. At one point, attempting to distract the raptors, Tim flings the baby toward them, hoping they will give up the pursuit to take care of it. Instead, they devour it, fighting among themselves for the scraps. But if the grown-up raptors are so fierce and smart and wicked, would they really have been such adorable cuties when they first emerged from the egg? Baby alligators and crocodiles, for example, come equipped with a full set of teeth and a bad attitude. They're hungry, and they'll try to eat whatever comes their way. Would a baby raptor nuzzle your hand or dart at one of your fingers and chomp it off?

When the first dinosaurs emerge in your high-tech hatchery, you'd be unwise to play with them as if they were chicks or bunny rabbits. Chances are you won't know what species of dinosaur you've got until it's hatched, and if you happen to have created a raptor or a *T. rex* or an allosaurus, or any of the numerous flesh-eating dinosaurs, your first thought might be to stay out of its way until you got an idea of how it was going to behave.

And in any case, might it not be sensible to see to it that the baby dino emerges into a secure and sterile environment? This is a creature adapted to live in a world that disappeared 65 million years ago or more. Undoubtedly there were diseases around in those times, but they weren't the diseases of today. A dinosaur hatching into the modern world is as vulnerable, in all likelihood, as the Indians of North America were when the first explorers arrived, bringing smallpox, flu, and other infectious diseases from Europe.

Then again, the question of keeping an infant animal germ-free is a tricky one. A young creature kept scrupulously away from all possible sources of disease from the moment of birth will, in all likelihood, fail to develop a fully functioning immune system, and will be all the more vulnerable as an adult. The case of the Indians in North America is instructive: adults were highly susceptible to the diseases brought in by the foreigners, but children, on the whole, did better. Those who survived went on to produce more children, who were then born into a world full of new diseases that they had to cope with from the start. And, in general, they coped as well as the children of the immigrants.

There's another, allied consideration. Our bodies are full of bacteria, viruses, and even tiny parasitic skin mites—most of them either harmless or actually beneficial. Our intestines, in particular, harbor a whole mini-ecology of microorganisms—bacteria, yeasts, single-celled protozoa—which has become essential to our digestion. A newborn human baby, like any other newborn creature, has none of that internal complement of microorganisms. But even though parents take some pains to keep their baby safe and clean and away from germs, the parents themselves are repositories of microbes, which quickly find their way into the new child. And that's a good thing.

We're hardly aware of all these internal microbes because they are ever-present. But in case you doubt their necessity, there's an unpleasant condition that can develop when they're absent. If you take a long course of antibiotics to get rid of one or another infection, sometimes you'll kill off all the bacteria in your gut except for a particularly hardy one called *Clostridium difficile*. When all the others are gone because of the antibiotics, the aptly named *C. difficile* comes out to play, and your intestines are its playing ground. If you have only this bacterium and no other in your insides, your digestion can't work the way it's supposed to and intestinal unpleasantness ensues. You then have to take a special antibiotic to beat back *C. difficile* and allow the usual bacteria to reestablish themselves so that your digestion can return to normal.

Amber-preserved insects, mites, and midges might provide some information on what sort of bacteria and other germs inhabited the Jurassic world, and that might be helpful in knowing how to keep the reconstructed dinosaurs healthy. But it would be impossible to reconstruct the entire Jurassic ecology down to every last virus, bacterium, and dust mite.

But perhaps that's not necessary anyway. There's good evidence that lots of insects and microorganisms have been around, in something pretty close to their modern form, for a long, long time. (Remember the bacterial spores that Raul Cano and Monica Borucki resurrected from an ancient amber-trapped bee? They seemed reasonably close cousins of the bacteria that live inside bees today.)

So maybe instead of fussing over your little baby dinosaur

and keeping it safe from every conceivable danger, you would be better off raising it in a reasonably clean but not surgically sterile environment. Inevitably, that means you're going to lose some number of baby dinosaurs to disease, but there's really not much you can do about that—especially if you want to run a park where dinosaurs are roaming free, in the wild.

To begin with, you have to figure out how to feed your little creature. If it's a vegetarian dinosaur, you can start it out on some plain mush of corn or oatmeal, maybe even some mushed-up prehistoric plants, if you've managed to cultivate any. But if it's a *T. rex*, it's going to want meat from the git-go. Maybe *T. rex*'s mother would feed it regurgitated bits of prey, the way most birds feed their young today. Or maybe she would just drop a live mouse into the nest and let junior catch his dinner. That's how the dastardly Lewis Dodgson finally meets his end in *The Lost World:* a mother *T. rex* catches him, but rather than munching him up there and then, she takes him back to the nest, where the screaming, struggling biotechnologist gets fed to two *T. rex* babies. The youngsters have a bit of trouble with Dodgson, who's trying to get out of the nest, so mama reaches over and nips one of his legs to break it. That way, the babies still have live prey to feed on, but the prey has been immobilized so the kids don't have to work too hard. Thus do mothers teach and encourage their children.

It won't be too difficult to discover whether your dinosaur is a meat eater or a herbivore. After all, paleontologists have generally been able to figure that out just from fossil remains— by studying the construction of the jaw and mouth and the shape of the teeth. Often, the physical appearance of the skeleton tells you whether the animal was built for hunting or for standing around in a field chewing grass.

Even if it's a species you don't recognize (or if the baby is sufficiently different from the adult that you can't quite be sure what it will grow into), you can tell from the conformation of its face and jaw what sort of things it's likely to eat. Then you can just give it a selection of appropriate foods and see what it goes for. If newborn babies of all species instinctively know only one thing, it's how to eat. For a herbivore, you can try milk, mashed potatoes, puréed carrots—all the usual kinds of supermarket baby foods. For a carnivore, maybe you'd throw in some raw hamburger or fresh-killed mice, as people do who keep lizards and snakes as pets. Or you might try insects such as crickets, which you can buy at pet food stores. Do keep your own fingers out of the way.

There are also specific indications of what dinosaurs might have eaten, because of the abundance of fossilized plants and insects and other small creatures from the Triassic, Jurassic, and Cretaceous periods. Paleontologists have been able to reconstruct in some detail what the landscape might have looked like, what sort of trees and plants were growing, and what kind of animals might have been scooting around at the dinosaurs' feet. One of the most informative sources on dinosaur feeding habits are coprolites, the polite scientific name for fossilized poop. By examining coprolites in minute detail (they're many millions of years old and turned entirely to rock, so there's no need to be squeamish about it), you can discover pollen grains, grass seeds, bone fragments—any number of remnants that can tell you what the creature that produced the poop had been eating the day before.

John Hammond has seen to it that the landscape of Jurassic Park has been planted with trees such as cycads, ginkgoes, and an order of primitive plants called Bennettitales, all of which were well represented in the Jurassic period. There's also a plant that Ellie Sattler notices early in the visit—a fern by the name of *Serenna veriformans*, which today is common only in Brazil and Colombia but was widespread long ago. Sattler knows that this plant, authentic as it might be, produces a toxic substance that makes its fronds irritating to the skin and potentially deadly when eaten. Who knows whether dinosaurs could have eaten it or not? Just because a plant was a genuine part of the Jurassic landscape doesn't mean that it ought to be planted in Jurassic Park. Many plants have evolved defense mechanisms against the animals that live in the same environment; filling Jurassic Park with plants known to have occupied dinosaur habitat might be exactly the wrong thing to do.

On the whole, though, you probably don't need to worry too much about providing specific foods for the dinosaurs. If the vegetarian dinosaurs ate leaves and grasses of the Jurassic period, they would probably do just fine on modern leaves and grasses, as long as you exclude anything you know to have toxic substances in it. (Of course, you don't know for sure what substances would have been toxic to dinosaurs in particular, but in general the poisons that plants produce belong to a smallish number of families of related chemical compounds, so you can more or less tell what's likely to have been dangerous to them.)

Likewise, if a meat-eating dinosaur was accustomed to chasing after little mammals that populated the Jurassic world, it would probably do just fine on lamb or goat or pig. Human beings, after all, have numerous preferences and taboos about what to eat (sheep are widely eaten, cats and dogs are generally not, horses are OK to eat if you're French, and so on), but these customs have little to do with nutrition per se. Like most other primates, humans are omnivorous—able to eat all kinds of things and willing to try most of them. Big dinosaurs, especially, would have had to eat a lot just to stay alive, so it's unlikely they were fussy about what they were chasing.

So food itself isn't too much of a problem, but at this stage, as you're trying to raise your baby dinosaur, you may be worried about disease and infection. The best bet is probably to feed your dinosaur using the same sort of hygienic precautions as you would for yourself or your own baby. Keep everything clean, don't feed the critter obviously old or bad food, but don't sterilize or pasteurize everything, because that would isolate the infant dinosaur from all the bugs and germs that it's going to have to learn how to deal with one way or another.

One thing you wouldn't do, at least to begin with, is feed the meat-eating dinosaurs mice that you caught in the wild. Such animals could harbor all kinds of diseases and parasites that you don't want your baby dino coming into contact with. But it's easy enough to breed mice in your dinosaur hatchery, and to keep them clean and away from the outside world. With luck, lab-raised mice will have the internal bacteria and other microorganisms that the dinosaur needs to acquire but won't have any devastating diseases it can't handle.

So you give your dinosaur meat that looks and smells and tastes fresh. But would it rush to eat the meat, or stare uncomprehendingly? Perhaps it's expecting to get some chewed-up, partially digested meat direct from its mother's mouth. Well, you've devoted a lot of time and effort to this project, and this is no time to get finicky. So you chew on some raw meat for a while and spit it out. And you breathe a sigh of relief as the infant begins to eat.

Or maybe it doesn't. Maybe it wants live food—worms or mice or scraps of still-twitching flesh torn by its mother from a struggling victim. Animal experts working at the Smithsonian Conservation Biology Center near Front Royal, Virginia, have discovered, after a lot of trial and error, that mynahs do much better at raising young in captivity if there's a wide variety and abundance of food available. At certain times, the mynahs like to feed live crickets to their infants; at other times, they prefer dried and prepared foods. It also seems that the presence of many different kinds of foods encourages the mynahs to mate and raise children in the first place, presumably because evolution has "taught" them not to have children when times are harsh—a propensity not unknown among humans, for that matter.

So you throw a live mouse into the dinosaur's cage and see what happens. A kitten will instinctively go after any scuttling little animal or insect, so perhaps your dinosaur will do the same. Of course, many kittens seem to regard mice and insects as especially entertaining toys, not food, but the dinosaur hasn't gone through thousands of years of domestication, and will probably catch and eat what you throw at it.

One of these strategies, you have to hope, will get the dinosaur's attention. First problem solved: it's eating and you've figured out what to feed it. That doesn't mean your difficulties are at an end, because for all you know there may be some strange mineral additive or vitamin the infant dinosaur needs. Humans contract a variety of odd ailments if they don't have enough vitamins—rickets, beriberi, and, of course, scurvy, the old affliction of sailors who didn't get vitamin C. That last one is a curious story. Somewhere along the line, humans, monkeys, and, even more curiously, guinea pigs lost the ability to make vitamin C, which is why we have to get it from what we eat.

But who knows what was a "vitamin" to a T. rex or a triceratops? How similar were all the different kinds of dinosaurs anyway, as far as their metabolism was concerned? And the so-called deficiency diseases can be enigmatic in their effects—it was basically a matter of trial and error that led a Scottish naval surgeon in 1753 to conclude that scurvy could be prevented by providing sailors with fresh fruit (limes were popular, hence "limeys" for English sailors); and even then no one knew specifically what it was in fruit that cured scurvy. The only consolation is that deficiency diseases tend to be slow-acting, which gives you some time to figure out what trace chemical is absent from your baby dinosaur's diet.

At any rate, your new charge is eating and growing, and you can turn your attention to other issues. Disease will probably be the biggest worry and the main cause of infant mortality. One of the memos that Malcolm and the others come across in the abandoned Site B mentions that many of the baby dinosaurs died after they were only a few days old. The cause, apparently, was contamination with *E. coli* bacteria, and the memo urges all technicians to take extra care to keep their egg-preparation procedures sterile.

Now, *E. coli* is such a common bacterium that you really can't hope to keep your young dinosaur away from it. What's more, *E. coli* may well be essential to its digestion, as it is to yours. A rare and virulent strain of *E. coli*, the notorious O157, recently caused some fatal cases of food poisoning in Japan, Scotland, and the Seattle area. But if you have to expose your dinosaur to the nice kinds of *E. coli* so that they can take up residence in the dinosaur's gut, how are you going to keep away the nasty strains?

To some extent, ruling out any obvious sources of bad food is just a matter of common sense. You might also want to protect the dinosaur from any of your lab technicians and assistants who have come down with a cold or other minor ailment; on the other hand, you can't give your dog a cold, and your dog can't give you distemper. Many bacteria and viruses have evolved in concert with one species or related group of species and cause no harm to others.

It's conceivable, however, that certain modern germs are part of a lineage that goes all the way back to the age of dinosaurs; and one or another modern microbe might still, as a sort of evolutionary accident, be able to infect the dinosaurs now revived in your lab. Some entirely innocuous bug that's everywhere in the human world and doesn't generally cause us much harm may turn out to be deadly to the dinosaurs. This scenario is the equivalent of the one that was played out early in this country's history, when much of the unprepared Indian population succumbed to relatively mild strains of influenza—strains that caused the Europeans only minor inconvenience. The additional possibilities for disaster that arise when you think about diseases or parasites crossing from one species to another are literally incalculable.

But, again, there's very little you can do to prevent such problems. Sooner or later, you're going to have to expose your new dinosaur to the outside world, with all the dangers, known and unknown, that that entails. Apparently, that's what the scientists ended up doing in *The Lost World*. Once the dinosaur young were past the immediate dangers of infancy, there seemed no alternative but to let them loose on Isla Sorna and allow nature to run its course—not that there's anything very natural about a lot of dinosaurs roaming across a Caribbean island.

There's not much you can do to give your dinosaur a guarantee of a long and healthy life. You can study its blood and respiration and stomach juices and so forth, but none of that will tell you what sort of diseases a dinosaur could have coped with in its original environment, what sort of diseases might have killed it, or how it will react to diseases of the modern world. Because of the way they have been re-created, the dinosaurs of *Jurassic Park* are at a distinct disadvantage. They carry none of the parasites, viruses, or bacteria they would normally have carried in their original world; they are immunologically pure and they are being dropped into a strange world full of creatures big and small that they have never encountered before. Human visitors will bring all kinds of germs to Jurassic Park—microbes that could cause fierce diseases in the innocent dinosaur population. You just have to hope those diseases will be more like tourist's tummy than some dinosaur version of Ebola fever.

You'd be wise to take things slowly as you introduced your baby dinosaur to its new world. You can give it water from a rain bucket or a stream, instead of the purified tap water it's been getting so far. You can try feeding it small animals captured in the park itself instead of giving it mice from your laboratory-bred supply. Or you might give it some leaves from a tree growing on the park grounds, rather than feeding it washed supermarket lettuce. You don't want to throw the whole modern world at it all at once: it might be overwhelmed by a slew of minor diseases that it could deal with one by one but not en masse. If you expose it to one or two things at a time and watch carefully for any signs of disease or ill health, you have a chance to protect it. (This is assuming, of course, that you can tell a healthy dinosaur from a sick one. If it groans and coughs and sneezes, you can be fairly sure something's amiss, but would you know if your dinosaur was grumbling with a stomach upset or growling with happiness? It's hard enough for parents to know if their colicky three-month-old human being is really sick or just having a bad day.)

There's another kind of problem that can afflict the dinosaurs. As Henry Wu remarks, he and his scientists sometimes found that "an animal grows for six months and then something untoward happens. And we realize there is some error." He means, specifically, an error in the genome that was used to create the animal. You'll recall that a good deal of guesswork went into the genome to fill in missing pieces or patch up disrupted segments, so it's no surprise that some of the genomes are not precisely what they're supposed to be. Whether it's likely that some random flaw in the genome would have this sort of subtle effect—damaging the dinosaur somehow but not enough to stop it from surviving the first few months of its life—is another question. An error can be small or large: as minor as a single base—a G instead of a T, say— somewhere in the billion or so letters that constitute the dinosaur genome, or a substantial sequence of bases missing or duplicated. What's surprising is that the seriousness of the consequences can be quite unrelated to the magnitude of the error itself. Sometimes a single base change can be fatal; other times, a whole chunk of one chromosome can be doubled up, and you might never notice the results. The human disorder known as cystic fibrosis, which is most common among Caucasians, afflicting about 1 in 10,000 babies, is the result of a single letter error in a single gene. Down's syndrome (so called after the doctor who first described it) is another fairly common genetic disorder; it includes some mental retardation and, like CF, tends to restrict the life span of its sufferers to about 30 years. Compared to CF, though, the error that creates Down's syndrome is huge: an entire chromosome, number 21, is repeated, so that a person has 3 copies instead of the usual 2. This enormous genetic error has, you might say, surprisingly limited consequences.

Then there's the condition known as fragile X syndrome. Along the X chromosome there's a triplet consisting of the DNA bases CAG, which is repeated, typically, 20 or 30 times. No one knows quite what this CAG repeat section does, if anything, and the precise number of repeats varies from person to person with no apparent significance. But in a few people, the X chromosome has many more of these CAG repeats, up to as many as 200 in some documented cases. When this happens, the consequences become prominent: as the number of repeats increases beyond about 50, people who carry the chromosome show increasing signs of mental retardation. The syndrome (called "fragile X" because the number of repeats makes the chromosome fragile and liable to break) affects only men, who have a single X chromosome. Women have two X's, so if they inherit a bad copy from one parent its ill effects are compensated for by a good copy from the other parent. It has been suggested that a large number of CAG repeats in a gene trips up the biochemical machinery that makes the gene's protein. This so-called transcription and translation mechanism runs smoothly over a short CAG repeat section but gets derailed somehow when the number of repeats is too big.

What this all boils down to is that the consequences of errors in reconstructed dinosaur genomes vary from total insignificance to fatality in an almost completely unpredictable way. This makes some of what Henry Wu says a little hard to believe. He implies on a number of occasions that once they've got the genome figured out and have managed to grow a dinosaur from it, they can tinker with little bits and pieces of the genetic code, just as you might clean up a few bugs in a piece of computer software. At one point, Wu comes to talk to Hammond bearing a file marked "Version 4.4" (as if it were the manual for the latest edition of a word-processing program). This file contains, evidently, Wu's redesigned genomes for a new generation of dinosaurs more to his liking. "We could easily breed slower, more domesticated dinosaurs," he tells Hammond. Hammond, predictably, wants none of this wishy-washy attitude and insists on "real" dinosaurs—although, as Wu points out, the dinosaurs they've already made can hardly be called real, since they contain bits of frog DNA.

Needless to say, Hammond wins the argument. But unless Wu is a whole lot more brilliant than anyone suspects, the argument is moot. There's no way to tinker with genomes to make dinosaurs more "domesticated." To be sure, evidence exists that some aspects of behavior have genetic roots. There's a mounting belief among scientists that human babies tend to be, for example, either shy or outgoing from the outset—that their fundamental attitude toward the external world is inbuilt. But a division into "shy" or "outgoing" represents only one element of many that contribute to personality. As babies grow into children and then adults, their behavior changes, depending not just on genetics but on the way they're raised. And parental behavior, in turn, depends on the child's attitude; parents will likely treat a shy child differently from an outgoing one, either reinforcing or trying to change their baby's attitude.

Likewise, some animals tend to be docile, while others are aggressive. Even after thousands of years of genetic engineering the old-fashioned way—by careful breeding through the generations—dogs and cats behave in fundamentally different ways. Some dog breeds are more people-friendly than others, some are said to be good with children, some are quick-tempered and snappish: nevertheless, all dogs tend to behave in certain doggy ways that cats do not. Breeding can alter an animal's behavioral characteristics somewhat, but a gregarious, group-loving animal like a dog can't be turned into a solitary creature like a cat, no matter how selectively you breed. If raptors really were mean, cold-eyed killers, Henry Wu would not have been able to tweak a couple of genes here and there to turn them into loving pussycats. If you wanted to make a quiet and peaceful Jurassic Park, where all the animals were kind and thoughtful and *T. rex* was happy to feed on fish sticks and hot dogs rather than tearing the flesh of its hapless victims, then you wouldn't have a dinosaur park at all but a barnyard version of Jurassic Park, bearing the same relationship to the real Jurassic period as Old Macdonald's farm does to the Serengeti.

Henry Wu's enthusiasm for genetic tinkering leads him into one last error. To make sure that the dinosaurs can't survive away from Jurassic Park, he alters their metabolism in a subtle way. He explains to the skeptical Malcolm that he's "made them lysine dependent. Unless they get a rich source of exogenous lysine, they'll go into a coma within twelve hours and expire."

Now, lysine is an amino acid (you can buy tablets of it at the health food store, and it's used as a supplement in chicken feed). It's an essential component of certain proteins. What Wu says he's done is to give all his dinosaurs a single faulty gene, so that they can't make lysine for themselves and have to get lysine supplements in their food, otherwise they'll die. Clever idea, don't you think?

Well, maybe not so clever. As a matter of fact, you and I can't make lysine either and have to get it from our diets. Of the 20 amino acids we need, we can make 11 within our bodies. The other 9, including lysine, we get from food. And despite the warnings of some health food fanatics, we can get enough of these 9 if we just eat a reasonably balanced diet. We don't need any supplements.

What's more, this is true of essentially all animals in the world today. None can make lysine. Of course, we don't know for sure how it was with dinosaurs, but the chances are that they, too, got their lysine from their diet. In that case, Wu can't have altered the gene with which they make lysine because there was no such gene in the first place.

OK, so instead maybe Wu genetically altered them so that they would be unable to digest lysine correctly. But that would have been a prodigious feat of genetic engineering. The network of biochemical reactions that extracts lysine from food, transports it within the body, inserts it into cells, and correctly incorporates it into proteins is fiendishly complex. Moreover, some of the steps involved also serve a number of other purposes, so if you disrupt the network you will cause more damage than you bargained for. And in any case, if you do prevent dinosaurs from properly absorbing lysine, how are you going to give it to them in a form they can use? With regular injections, like insulin for diabetics? Even the rugged Muldoon might balk at the idea of visiting tyrannosaurs with a giant needle every day to give them their shots.

For all Wu's remarks about "fine-tuning" the biology and behavior of his dinosaurs, the only way to find out how they will behave in the wild—and, indeed, whether they will survive there at all—is to throw them out of the lab and hope for the best. As it is with raising children, there's only so much a parent can do.

CHAPTER EIGHT

FINDING A HOME

You've raised a clutch of dinosaurs from eggs, figured out what to feed them, and kept them safe and warm and apparently free of disease while they're growing into juveniles and young adults. Now comes the time when you proudly let them loose to romp through the spacious parklands and tangled forests you've prepared for them. Once out in the open, they seem to know instinctively where to live, what to eat, how to survive. Pretty soon, you have an authentic-looking prehistoric world, inhabited by dinosaurs behaving in the wild as they did many millions of years ago.

Well, maybe. Another possibility is that you let your dinosaurs loose and not long afterward they all starve or die of exposure because these orphans haven't learned what to eat or where to hunker down for the night.

Then again, most animals have a pretty strong urge to survive. They're not likely to starve to death in the midst of plenty. Instead, they eat up a storm. The plant-eating dinosaurs strip the cycads and the gingkoes and trample the grasslands into mud. The meat eaters ambush and devour the docile plant eaters, who don't appear to know enough to be afraid of *T. Rex* and *Allosaurus*. They stand around in the open while the carnivores pick them off, just as the dodos on the island of Mauritius stood around clueless while British sailors clubbed the whole lot of them to death.

In short order, all the animals either get eaten up or run

145

out of food. They die, having laid waste to the lavish park you set out for them.

Your young dinosaurs represent billions of dollars and many years of effort. You can't just let them all loose in a field and hope they'll figure out how to survive. There's no one except you to teach them how to obtain their own food, where to hide, what to chase, what to run away from. Somehow, you're going to have to teach the dinosaurs those living skills yourself—and, like any anxious parent, you have to hope that the kids don't start suspecting that you don't really know what you're doing.

Your dilemma is this: your young dinosaurs will be exposed to human contact, because they have to be fed and looked after; but you also want to protect them from human contact as much as possible, because you want them to grow up as dinosaurs and not pets, or confused creatures that aren't quite sure what they are. In the 1930s, the pioneering German biologist Konrad Lorenz discovered that baby geese would, at a certain time in their lives, fix upon their caretaker as their "mother." Lorenz himself became the designated mother of a flock of goslings, and once this "imprinting" had happened, the creatures dutifully followed him wherever he led. These geese weren't very good at being wild geese; moreover, once they'd imprinted on Lorenz it was impossible to get them to recognize and follow a goose instead.

How will your dinosaurs react to being raised by humans? The answer will undoubtedly vary from species to species. You might think the answer is to raise the animals without *direct* human contact—they could be housed out of sight of people, for instance, and fed by machines—but that won't work either, because the youngsters will get no guidance at all on how they should behave. Although some forms of activity seem to be hardwired into the brain, most animals learn behavior and survival skills as they grow and develop within a preexisting community—a working system of adults, adolescents, and juveniles. With guidance from parents, and perhaps with discipline from other members of the animal community, young creatures "learn their place."

The "nursery" for the dinosaurs of Jurassic Park turns out to be the secret island—Isla Sorna, of *The Lost World*. There Henry Wu and his team, because they can't figure out how to raise the dinosaurs they've created, more or less give up the effort and let the animals run wild. A sort of hastily enforced natural selection takes place: those dinosaurs that manage to survive are transported to Isla Nublar—Jurassic Park itself—to be part of the exhibit; those that don't—well, too bad for them. There are always more dinos where those came from.

The dinosaurs of Site B display a variety of dysfunctional habits. Ian Malcolm attributes the nastiness of the velociraptors to lack of parental guidance: raised without any constraints, they become reckless carnivores, chasing prey at every opportunity and not always eating what they catch. Crichton depicts the velociraptors' lair as something out of the movie *Animal House*—Bluto Blutarski's frat room on a bad day—with half-eaten scraps of food scattered everywhere and raptors bedding down for the night wherever they can push each other aside to make a space. What's more, the raptors are mean to their own offspring. When a pack of raptors has brought down a large animal, the whole herd descends on the corpse in a feeding frenzy, and young raptors have to fight their parents for scraps. The zoologist Sarah Harding notes that this behavior is in marked contrast to that of packs of hyenas on the African plains: hyena parents make room for the youngsters at the kill and see to it that everyone gets a bite to eat; indeed, it's common for most predator groups in the wild to develop a "pecking order," according to which all group members get their chance at food.

But if the velociraptors have so dismally failed to learn any social niceties, why have the triceratops, apatosaurs, and parasaurs done any better? The plant-eating triceratops of Site B turn out to be admirable parents. When the triceratops herd is under threat, the adults form a ring around the youngsters, protecting them. The apatosaurs have even worked out cooperative behavior with another species—their fellow vegetarians, the crested parasaurs. The towering apatosaurs can defend themselves with a swipe of their heavy tails, but they have poor eyesight; the parasaurs, on the other hand, are sharp-sighted but smaller and more vulnerable. So the two herds intermingle, the parasaurs providing the warning signals and the apatosaurs acting as the defense. The parasaurs have evidently managed to acquire even more sophisticated patterns of behavior. In one instance, Richard Levine watches in puzzlement as two parasaurs, making strange honking noises, walk away from the herd. The rest follow, and, eventually, they all arrive at a communal "latrine" area some distance from the meadow where they have been grazing.

Such behaviors—cooperating with another species, herding to protect the young, following troop leaders to a latrine—may well have genetic roots, but even so, whether or not those characteristics develop depends a great deal on the actions of parents and on the general environment. Behavior in animals, as well as in humans, emerges as an inseparable mix of innate and learned characteristics. Orphan dinosaurs raised in the lab may display some sort of innate social behaviors when they are released into the wild, but there's no telling whether those behaviors will cohere into a social structure, or whether that structure will at all resemble that of a true dinosaur community.

In *Jurassic Park*, incidentally, all the animals are said to be females—a precaution taken to prevent the herds from reproducing. In *The Lost World*, the herds obviously consist of males and females, since they have been living and breeding for 5 years before Malcolm, Harding, and Levine discover Isla Sorna. It's not clear whether some of the *Lost World* dinosaurs changed from female to male—as they did on Isla Nublar, for the dubious reason that bits of frog DNA in their genomes made them transsexual—or whether Henry Wu and his team had been recklessly producing males and females all along. In any case, a purely female (or purely male, for that matter) herd of dinosaurs would be a strange and unpredictable thing, since sexual competition sparks cooperative or aggressive behavior in many animal species.

The velociraptors of *Jurassic Park* and *The Lost World* are an odd case. They're greedy, selfish, and aggressive—aggressive toward each other and even toward their own infants. On the other hand, they're also skillful pack hunters and appear to put their rivalries aside while chasing their prey: one raptor will maneuver a weak, straying triceratops away from its herd while another waits in hiding to pounce on the prey. But how did orphan raptors acquire the cooperative skills needed for hunting while remaining completely devoid of the similarly cooperative skills needed for living and surviving in a pack? In particular, the idea that raptors would deny food to their own infants is difficult to believe. Most animals display an instinctive protectiveness toward their offspring—their own offspring, at least. Without this instinct, no species is likely to survive. Who knows what basic survival skills mother dinosaurs, or a herd of adult dinosaurs, would have taught dinosaur youngsters? Surprisingly, paleontologists have been able to make a few simple inferences about the way some dinosaurs raised their young. Jack Horner, uncovering the nest sites of great herds of maiasaurs on the North American plains, discovered bones of juveniles, up to a moderate size, either at or near the nests. It seems reasonable to guess that the maiasaur parents kept their offspring near the nest and fed them until the youngsters reached the point of self-sufficiency, rather than abandoning them as soon as they were hatched. But did both parents look after the children, or just the mothers? Did parents look after only their own children, or did they form cooperative nursing groups? Those questions are harder to answer.

Similarly, the paleontologist Robert Bakker has uncovered fossilized "trackways" in which the footprints of apatosaurs are preserved. He sees bigger prints toward the front of the herd and smaller ones behind them, suggesting that the adults were protecting the infants. In *The Lost World*, triceratops act in a similar way. But did dinosaur herds behave like this all the time, or only when traveling, or only when they perceived danger? Hard to know.

The plain fact is that even if some sort of stable dinosaur society does emerge, you won't really know whether your mini-ecosystem reflects the way real dinosaurs would have lived tens of millions of years ago. Moreover, you're putting together a variety of dinosaurs from different eras—dinosaurs that missed each other by many millions of years—and you're forcing them to live in a confined space. In reality, some of those creatures that were contemporaries may have learned to stay well out of one another's way. In Jurassic Park and on Site B, they don't have that option.

Since you have to put the dinosaurs in a limited space, the first thing you want to be sure about is that it's enough space. Isla Nublar measures 8 miles long by 3 miles at its widest point. John Hammond, describing Jurassic Park to his guests, notes that the area of Isla Nublar is 22 square miles—but there's obviously a certain amount of space taken up by the employees' quarters, the visitors' center, the research labs, and so on. How much space is left for the dinosaurs themselves? Let's be generous and suppose that some 20 square miles are open for the dinosaurs to roam in. That's equivalent to an enclosure about 4.5 miles on a side. Sounds pretty big, you might think—but then some of the dinosaurs are pretty big, too, and will need a lot of space.

In Montana, as the novel opens, Alan Grant and Ellie Sattler are digging up dinosaur fossils in an area where they have found almost nothing except hadrosaurs, large grazing animals. They estimate, from the number of remains and the way they're scattered over a large area, that the hadrosaurs roamed the ancient North American grasslands in herds of 10,000 or 20,000, as buffalo would roam the same area millions of years later and wildebeest still cascade across the African plains today. The reason Grant and Sattler find endless numbers of hadrosaurs but little else, they decide, is because predators must have been much rarer: a herd of 10,000 hadrosaurs would have "supported" a population of about 25 tyrannosaurs. More than that and the hungry meat eaters would have thinned out the hadrosaur population faster than they could reproduce.

Other dinosaurs, notably the raptors, seem to have had the physiques of fast runners. In the modern world, cheetahs can reach 60 miles per hour, and plenty of animals can sustain 30 mph for extended periods. Dinosaurs capable of running at such speeds could have sprinted from one end of Isla Nublar to the other in a matter of minutes, which they must have found frustrating in the extreme.

There's not nearly enough room in Hammond's Jurassic Park for 10,000 hadrosaurs; you'd be lucky to raise a population of 100 large grazing animals in so small a space. But 100 hadrosaurs—according to Grant and Sattler's estimates—would support only $^1/_4$ of a *T. rex*—which, needless to say, won't exactly form a self-sustaining population. Looking at things the other way around, you might decide that you need at least a couple of dozen tyrannosaurs to form a viable tyrannosaur population; any less than that and passing bad luck—an outbreak of disease, say, or a year in which all the offspring died young—might wipe the lot of them out. But to sustain that many tyrannosaurs, you'd need tens of thousands of hadrosaurs, and for that you'd need a Jurassic Park the size of Connecticut, not some poky little island in the middle of nowhere.

Hammond and his team attempt to get around this problem by making Jurassic Park into what is basically a zoo. The animals are confined to different areas (Velociraptor Valley, Carnivore Country, and the like), and the carnivores are fed. This is not a functioning ecosystem, even in an artificial way. Animals need places where they can hide and sleep and breed. The tally of animals on the island includes 2 tyrannosaurs, 8 raptors, and 7 dilophosaurs (an early Jurassic carnivore). That's 17 voracious meat eaters in a population of 238 animals altogether. Confining the carnivores is an excellent idea, because left to their own devices they would undoubtedly polish off the other animals in a matter of weeks and then fight it out with each other. The winners of the battle for supremacy between raptors and tyrannosaurs would be rewarded by starving to death, since there would be nothing left to eat. But if you're feeding the animals rather than letting them get their own meals, you've got a lot of feeding to do. In one of the movie's most riveting scenes, a hapless goat is pegged to a stake in the *T. rex* compound, and the poor thing bleats pitifully as the ground begins to shake under *T. rex*'s approaching footsteps. Pretty soon, no goat.

How many goats a day would constitute a healthy diet for

a full-grown *T. rex*? Here's a chance to do what scientists like to call a back-of-the-envelope calculation. A person weighing 200 pounds will eat a few pounds of food every day. (Think of how big a plate of food you'd get by putting together all your meals and snacks for one day.) So let's say a typical animal needs to consume about 2 percent of its own body weight every day—4 pounds of food for our 200-pound person. (The figure would be somewhat higher if dinosaurs were warm-blooded and lower if they were cold-blooded.) *T. rex* is thought to have weighed something like 7 tons, which is around 15,000 pounds; 2 percent of that is 300 pounds. How much does a full-grown goat weigh? About the same as a medium to big dog, perhaps; let's say 75 pounds. It looks as if we need 4 goats a day to keep Mr. *T. rex* happy.

OK, 2 tyrannosaurs, 8 goats a day. The raptors are smaller—about 6 feet tall and relatively slender, but also well-muscled and powerful. Let's guess that they weighed about 500 pounds each, in which case 2 goats a day could keep 8 raptors well fed. Then there are the 7 dilophosaurs—10 feet tall, a bit bigger than raptors. So let's throw in at least another couple of goats for them.

You can see that feeding these carnivorous dinosaurs is not going to be a piece of cake, so to speak. Altogether, you're looking at a minimum of a dozen goats a day. If you had a goat ranch on the island, you'd need a substantial herd to be able to remove a dozen every day. You might need as many as 1,000 goats. That doesn't seem to be a good solution when you're squeezed for space, so you'd have to ship the goats in. But Hammond mentions to his visitors that deliveries by sea are unreliable because of the weather. You can't count on daily goat deliveries, so you'd have to have some sort of pen on the island in which you could keep several dozen goats in case the ship didn't get through for a few days. And each ship that arrived would have to deliver several dozen goats, to replenish the larder. And you'd have to feed all those goats, too, while they were waiting to be fed to the dinos. . . .

Maybe things won't be so difficult where the plant eaters are concerned, though. At least they can graze in the meadows of Isla Nublar. Unfortunately, in his desire to impress visitors, Hammond has put 17 apatosaurs in the park. Now, these Jurassic creatures weighed 20 to 30 tons apiece, which is roughly equivalent to about 50 good-sized dairy cows: therefore, you have the equivalent of almost 1,000 cows grazing on your island. You could sustain a herd of 10 or 20 cows on a few acres of grass, depending on how lush it is; a square mile is 640 acres, so the herd of 17 apatosaurs will need about a square mile all to itself, at the minimum. Then there are 21 maiasaurs, according to the park's computer tally. These animals were some 30 feet tall and must have weighed 10 tons at least. (By comparison, an African elephant weighs about 5 tons.) So the maiasaurs represent another several hundred cow-equivalents—almost half a square mile for the maiasaurs, therefore. And there are 11 hadrosaurs, 33 hypsilophodontids, 8 triceratops, and assorted smaller herbivores.

It's starting to look as if the whole of Isla Nublar is getting taken up with meadows and grasslands for the plant eaters, and we've still got to find room for the carnivores and the goat farm to supply them. And let's not forget that the island is described as a "volcanic upthrusting of rock," a "rugged and craggy" place with forested slopes. Sounds as if a good part of the island would not be happy terrain for the maiasaurs and apatosaurs, which like grassy plains, not forests or craggy slopes.

So it might be that you would have to bring in grass and hay too, along with the goats. This is getting difficult. You're going to need a pretty big ferry coming over every couple of days with a small herd of goats and some impressive bales of hay for the apatosaurs and maiasaurs, the hadrosaurs and hypsilophodontids.

*S*ite B, the Lost World of Isla Sorna, is a different proposition. There, things have been left to go wild for several years by the time the good guys discover the island, and when they arrive they find a lush place teeming with life and large populations of dinosaurs apparently doing pretty well for themselves. What's more, the island itself seems to be in good shape, with plenty of thick vegetation, trees, and meadows.

In *The Lost World*, Michael Crichton unfortunately didn't pass on to us the detailed memos and observation notes that paleontologist Richard Levine must surely have made. He spots herds of apatosaurs, maiasaurs, triceratops, parasaurs, hypsilophodontids, and other herbivores. Toward the end of the book, when Sarah Harding is trying to restart a broken-down vehicle, a herd of some 50 pachycephalosaurs, each about 6 feet tall, puts in an appearance. In the velociraptors' disheveled lair are 4 apatosaur skeletons, the remains of animals that died and were carried to raptor territory by a river, then eaten by raptors. All in all, Ian Malcolm estimates that there are a couple of hundred animals on the island, maybe as many as 500. It's unclear whether he's simply estimating the numbers of big animals here; there are also numerous chicken-size "compys" and rat-size mussaurs. (The case of the mussaurs illustrates one of the dangers of reviving dinosaur species without knowing exactly what you're doing. Originally discovered as fossil skeletons less than a foot long, this creature was named *Mussaurus*—"mouse lizard." Recently it was realized that the skeletons were those of infants and that the adults grew to a length of more than 10 feet! If Henry Wu had counted on the mussaurs' nicely complementing the scavenging activities of the compys—which are genuinely small dinosaurs—he must have been unpleasantly surprised when the little beasts just kept growing and growing.)

Apparently there are as many, if not more, animals on Isla Sorna as there were on Isla Nublar. The dimensions of Site B are not explicitly given, but judging by a map in the book and the distance the protagonists travel in their Jeeps, it's somewhat bigger than Isla Nublar. Let's say 100-plus square miles. Even so, since we've established that Isla Nublar was nowhere near big enough for the animals on it to form self-sufficient populations, it's unlikely that Isla Sorna is big enough for its dinosaurs to have lasted 5 years without killing each other off or (at the very least) destroying the landscape and vegetation.

There are also too many raptors—not to mention too many tyrannosaurs—on Isla Sorna for the few hundred other dinosaurs to support. This is a point that puzzles Malcolm and Harding. They count 12 raptors, a population that should have made much bigger inroads into the herbivore population than they seem to have done. The explanation can't be that there were a whole lot more herbivores at one time, because the herbivore population is already as large as the island can bear. What Harding and Malcolm decide is that the raptors suffer from an epidemic disease—a version of bovine spongiform encephalopathy, better known as Mad Cow disease. BSE has devastated cattle herds in Britain (perhaps more because of public opinion than because of the disease itself: as one indiscreet member of parliament put it, "It's not the cows that are mad, it's the journalists"). The raptors were originally fed sheep extract that contained the protein fragments called prions, thought to be responsible for BSE and other such diseases. Sheep suffer from a related ailment, called scrapie, and in humans the slow but fatal brain damage characteristic of Creutzfeld-Jakob disease seems to have a similar origin. The BSE-like disease spread through the raptor population, causing them to die young. "The rapid die-off supports a much larger predator population than you'd expect," Harding explains.

But this explanation makes no sense at all. It may be that the raptors die young, but when the visitors reach the island, there are 12 adult raptors on it. These raptors are old enough to breed, and breed they do: the nest site contains not just eggs but baby raptors. Twelve adult raptors is still too many: they would devastate the herbivore herds no matter what their life expectancy happens to be. The only way the Mad Raptor disease explanation would work is if there were never more than two or three full-grown beasts on the island at a time.

I f you want a place where you can set a representative collection of dinosaurs free and have them live as they once did in the wild, you need a considerably bigger island even than Isla Sorna. Where are you going to find one?

Apart from space, you need to worry about climate. It's not a matter of finding a climate that all the dinosaurs will like, because the dinosaurs that Wu raises for Jurassic Park didn't all live in the same conditions. As has been noted, they didn't even all live in the same geological period. More significantly, the climate in which these dinosaurs lived was substantially different from one period to the next. During the latest period—the Cretaceous, when *T. rex*, the raptors, the maiasaurs, and the triceratops held sway—the plains of North America, where many of these dinosaurs are fossilized, had a temperate climate, comparable to the climate there today. But during the Jurassic and Triassic periods (allosaurs, procompsognathids,

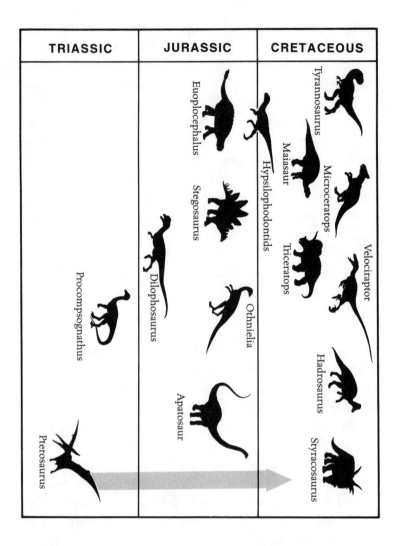

TRIASSIC	JURASSIC	CRETACEOUS

Euoplocephalus

Tyrannosaurus

Maiasaur

Hypsilophodontids

Microceratops

Stegosaurus

Triceratops

Velociraptor

Procompsognathus

Dilophosaurus

Othnielia

Hadrosaurus

Apatosaur

Styracosaurus

Pterosaurus

Timeline showing the periods of the various dinosaur species that were "cloned" and released on Isla Nublar. Note that most species are from either the late Cretaceous or the late Jurassic.

(*Drawing by Edward Heck*)

apatosaurs, hypsilophodontids), the climate was a good deal more humid and the environment was more like a steamy swamp.

That makes it tough to find one place, especially a single island, that can accommodate all of these creatures. It might be that they could adjust to an unfamiliar climate, just as polar bears at the zoo seem to do—but that's just one more way that Jurassic Park differs from a realistic dinosaur world. Michael Crichton describes both Isla Nublar and Isla Sorna as being steamy in places, because of underground volcanic heat, but it is also noted that the primary vegetation of Jurassic Park is that of a deciduous rain forest, making it sound something like the Pacific Northwest. Spielberg's movie depicts large areas of temperate grassland on Isla Nublar, along with denser jungles and swamps. It's an advantage of the movies that you can put all these dissimilar habitats onto one not very big island.

Islands are hard to come by these days. Anything substantial has been explored, mapped, colonized, or otherwise taken over by one country or another. It's hard to say just how big your island would have to be to accommodate a reasonable number of dinosaurs, but if you aspire to at least a couple of hundred square miles, you're getting into a conspicuously large chunk of real estate—Barbados, say, at 166 square miles, or Martinique, also in the Caribbean, just over 400 square miles. Buying an unknown, uninhabited island off the coast of Costa Rica is one thing, but taking over a sovereign country or a French département already occupied by the resort homes of millionaires is another matter altogether. Incidentally, Hammond's choice of Costa Rica may not have been such a smart move, because the Costa Ricans have been in the forefront of efforts to save their country from the predations of developers and uncontrolled settlement. It seems unlikely that Costa Rica would have allowed him to take over an entire island or two for unspecified high-tech purposes without asking for some sort of oversight of what he was up to.

You might explore a somewhat more remote part of the globe, perhaps in the South Pacific, but there, too, it's not easy to find an island big enough for your purposes. If you want 100 square miles or more, there are islands or island groups like Tonga, Fiji, Tahiti—or perhaps one of the Hawaiian chain. An island is a good idea, for security reasons, but the chances of even John Hammond being able to purchase one big enough and in a part of the world with a suitable climate seems unlikely. That leaves a mainland dinosaur park, with all the additional risks that that would entail. You could perhaps get hold of a large tract of land in Africa, South America, or in southern stretches of the former Soviet Union or Asia, but then you have to think about installing giant moats or electric fences or enormous walls to keep the animals in. A 30-ton apatosaur, even if it's in a benign mood, could do a lot of damage just by blundering cross-country.

There's another factor we mustn't overlook. Any piece of land suitable for dinosaurs to inhabit will already have a full complement of modern wildlife on it—not just plants, insects, and bacteria, but birds, lizards, rats, monkeys, rabbits, who knows what. It all depends where the place is.

Do you scour every large animal from the park before you ship in the dinosaurs, and hope that the dinosaurs themselves will establish their own replacement ecosystem? That's unlikely to work: even when dinosaurs dominated the earth, they weren't the only creatures extant. The earliest mammals date back to the Triassic, the dawn of the dinosaurs, and there were plenty of them around—small ones, to be sure; none bigger than a turkey or a small dog—throughout the Triassic, Jurassic, and Cretaceous periods. Who knows to what extent the dinosaurs depended on some of these creatures? No doubt a few of the smaller mammals were scavengers; without them, any dinosaur community would have been a mess. On Isla Nublar and Isla Sorna, the compys play the role of clean-up crew, taking care of the big dung piles the herbivores leave behind. There were plenty of other small dinosaur species, and small lizards and snakes besides. Smaller animals are as important a part of an ecosystem as the larger ones that reside at the top of the food chain. In the modern world, their abundance more than makes up for their size: there's much more animal mass in the form of rats, mice, and termites than there is in elephants and lions. Without these small but numerous creatures scurrying across every corner of the planet, we'd be up to our necks in waste and garbage in no time at all.

If you wanted to re-create a true dinosaur landscape, you'd have to find DNA for every single contemporary creature and make copies of those animals too. That's not going to be possible, as a practical matter (particularly given the whopping anachronisms you've already committed by re-creating dinosaur species originally separated from one another by millions of years). So the best you can do is allow the wildlife that's already in your dinosaur preserve to stay there, and hope that it can establish some sort of modus vivendi with the dinosaurs. (This is evidently what must have happened at Site B, although little mention is made of any non-dinosaurs on the island.) But what will result is an even less authentic ecosystem than we were already anticipating. Ian Malcolm's idea of studying Site B to discover clues to true dinosaur behavior, and possibly some of the factors that led to their extinction, is a mighty shaky one.

We simply don't know, in any precise sense, how dinosaurs moved, what sort of mating displays they might have made, whether they curled up to sleep like cats or hunkered down like cows, and so on. This gives Michael Crichton a good deal of license to invent convincingly realistic scenes of animal life on the two islands. One curiosity is that in *Jurassic Park*, Alan Grant is cornered by a *T. rex* but manages to escape harm by remaining absolutely motionless. Grant freezes not just because he's petrified with fear but because he believes that a *T. rex* can't see him as long as he doesn't move. The dinosaur roars and bellows—is it trying to provoke Grant into giving away his position?—but Grant stays stock still and eventually *T. rex* stomps off in frustration. What saves Grant, supposedly, is that *T. rex* has a vision system somewhat like that of a frog, which is unable to pick out stationary objects against a stationary background and relies on detecting movement.

What sort of vision *T. rex* really had is anyone's guess. It was a hunter, so presumably it must have had reasonably sharp sight, but hunters also work by sound and smell. In any case, in *The Lost World*, George Baselton, the Stanford biology professor and Lewis Dodgson's sidekick at the Biosyn Corporation, tries the same trick that Grant used to get away from a *T. rex*, and he fails dismally. He's trying to steal an egg from a *T. rex* nest when the noisemaking device he uses to scare the adults away abruptly fails. Baselton stands as still as he can, but the *T. rex* gobbles him up anyway. Richard Levine, who's watching this scene through one of the surveillance cameras scattered about the island, remarks rather contemptuously that it was a silly idea in the first place to think that an efficient hunter like *T. rex* could not have detected motion. Animals often freeze when attacked (as in the cliché about a deer in a car's headlights), and if this simple stratagem had fooled *T. rex*, he would have been one short-lived predator instead of the Curse of the Cretaceous.

Finally, we should not forget that one of the great attractions of Jurassic Park and Site B would be the range of sights, smells, and sounds that fill both places. The great charm of the books, and of Spielberg's moviemaking, lies in the fact that the fossilized skeletons of dinosaurs have been given flesh and colors and noises. These dinosaurs are green and brown and red and yellow; they honk and roar and trumpet; they trample noisily or slink silently; they have bad breath, runny noses, poisonous spit; they leave warm, steaming, odorous dung piles; *T. rex* even marks one of the Jeeps with white goo from its scent glands as a way of indicating ownership of that piece of territory—horrific enough, but at least he apparently doesn't mark his territory the way a dog does.

Most of this is reasonable in a broad way. Dinosaurs were highly evolved animals, and no doubt their colors, utterances, and behaviors rivaled the animal displays we're familiar with in the modern world. But if we want scientific evidence for the way any particular dinosaur looked, smelled, or sounded, we draw almost a complete blank. As to what colors the hides of various species might have been, what sorts of howls or grunts they might have produced—you might just as well make it all up.

Some scientists have tried to find out about dinosaur sounds by modeling the bony cavities of their heads and crests as if these were musical instruments resonating with a particular range of sounds and frequencies. That's not unreasonable, but the variety of squawks, honks, roars, and bellows that animals make depends much more on the soft tissue of tongue and throat than on jaw shape or skull cavity. Beluga whales have a fatty blob in their foreheads that seems to be an important part of their sound production system, but no one has quite figured out how it works. If you were given the skeleton of a cat, would you guess that it could make the noise we call purring? Basically, you can have some confidence that big animals make louder and deeper noises than little animals, but that's about it.

As for skin color, we know exactly nothing. Traditionally, dinosaurs have been portrayed in browns and greens, because (traditionally) they were thought of as large reptiles. But if you think of them as precursors to birds, a whole new range of colors comes to mind. The chameleonlike carnosaur of Site B, able to change the color of its skin at will and disguise itself as a chain-link fence, may be the product of a heightened imagination, but apart from that, why not give your fictional and cinematic dinosaurs bright colors and flabbergasting noises? Until someone creates a real, live dinosaur, nobody can prove you wrong.

CHAPTER NINE

THE MATHEMATICS OF MURPHY'S LAW

Dressed in black and constantly warning of disaster, the mathematician Ian Malcolm is perhaps meant to represent the conscience of Jurassic Park, the person who's always asking whether John Hammond and his genetic engineers are doing the right thing. As always happens in books and movies of this sort, nobody pays much attention to him until disaster strikes. But even after the fences fail and the dinosaurs break loose, Hammond, the true believer, continues to insist that with a better design and a bit more tinkering everything could have been saved. In the book, Hammond ends up getting eaten by compys, whose narcotic bites put him into a calm reverie. As he's nibbled to death, he muses to himself that things are working just as they should—the compys are, after all, doing exactly what they were meant to do. In the movie, though, Hammond is rescued. As the other survivors help him into the waiting helicopter, he takes one last wistful look at what remains of his creation, perhaps wondering how to build a new and improved park where all those accidents won't happen.

In its broad outline, at least, the movie is your standard thriller: the good guys triumph; the bad guys and the monsters are defeated. In the book, however, Michael Crichton conveys a much stronger message—about the tendency of scientists to do things just for the sake of doing them, and about the parallel tendency of things to go wrong because reality turns out to be more complicated, more difficult, and more unpredictable than scientists ever bargain for. Ian Malcolm, apparently Crichton's spokesman in the novel, argues constantly—to the point of making a nuisance of himself—that he knows for a fact that disaster is bound to strike.

As a plot device, the notion that scientists bite off more than they can chew and that unexpected but, in retrospect, obvious problems thereby crop up is a staple of the standard Frankenstein story. But in *Jurassic Park*, Frankenstein's monster—the menagerie of re-created dinosaurs—is not in itself responsible for the ensuing disaster. What's ultimately responsible is chaos theory. According to Malcolm, it's not just that things go wrong but that chaos theory—the mathematical theory of complex systems, to give it a more formal title—actually predicts that things will go wrong.

Chaos theory concerns the inherent unpredictability that can arise in complicated systems, such as you might get when, say, you throw a bunch of dinosaurs together under novel circumstances in a time and a place alien to them. The failure of the Jurassic Park experiment is meant to illustrate Malcolm's argument: no matter how much the scientists think they know, he predicts from the outset that the project will fail for unanticipated reasons. Malcolm can't predict exactly what will go wrong, but (according to him) chaos theory proves that something will. Chaos theory is made to sound like a sort of universal theory of failure.

Malcolm dresses up his dire warnings with a lot of fancy mathematics and obscure jargon, but the basic idea is one we all know by another, less elevated name. It's the well-known principle called Murphy's law: If something can go wrong, it will. The toast will always land butter-side down. If you leave your umbrella at home, it's certain to rain. Surely, you might think, there must be more to chaos theory than a fancy-dress version of your grandparents' folk wisdom!

Well, there is, of course, but it's not clear from the book just how much Malcolm really knows about chaos theory—or about elementary mathematics, for that matter. Or if he does understand such things, he's careful not to explain them in any depth to anyone else, so the aura of mystery remains.

For example, when the visitors are shown some of the computer systems that keep track of Jurassic Park's dinosaurs, they get to see a graph of dinosaur infant mortality during the previous 18 months of the park's existence. The numbers spike up and down from month to month, and Malcolm remarks that this irregular alternation is "characteristic of many complex systems." The idea he's hinting at here is that in systems where a great many variable factors influence an outcome—whether it's dinosaur mortality or the rate at which water drips from a faucet—complexity results in unpredictability. Instead of a nice steady drip, drip, drip (as Malcolm puts it), you get an irregular sequence of little drips and big drips, with seemingly random pauses in between. Another familiar example is waves crashing on a beach: they come in fairly regularly, but there's also a good deal of unpredictability about the timing and size of each individual wave. This approximate regularity has inspired all kinds of folk wisdom about subtle features of the surf—every seventh wave is big, according to one version—but mathematicians know that no simple formula can capture every aspect of wave behavior. That's a hallmark of chaos: As you try to describe some phenomenon in increasing detail and with increasing accuracy, you find that irregularities continue to crop up at finer and finer levels.

In the same way (Malcolm wants us to believe), the irregular pattern of dinosaur infant mortality is the result of a complex, chaotic system at work.

But if Malcolm tried to make this kind of argument in a scientific seminar, he'd be laughed out of the room. The problem is that the numbers are too small to be "characteristic" of anything. The variation depicted can easily come about by pure statistical chance. If you have a small population of dinosaurs and an infant mortality rate that, on average, is constant, you are bound to see variations from month to month. If 2 or 3 dinosaurs die in one 4-week period, and 5 or 6 in another, that tells you nothing at all.

Malcolm makes an even more elementary mathematical error. The computer that tallies the animals in the park also keeps track of their sizes, and it presents to Malcolm and the others a graph showing the distribution of heights among the procompsognathids: some of the compys are smaller than average, some larger, but most of them cluster around the middle of the height distribution, giving the standard bell curve. Malcolm then makes a correct observation: since the lab-created compys emerged in 3 separate batches at 6-month intervals, you wouldn't expect a bell curve for the population as a whole. Malcolm indicates the sort of three-humped distribution of heights that you *would* expect. Because there are, in effect, 3 distinct populations of compys of different ages, and therefore different sizes, each population would be characterized by its own bell curve. And Malcolm deduces that the reason the computer detects a true bell curve instead of the three humps is because the 3 compy populations have begun to interbreed, blurring the distinction between the groups.

That part of Malcolm's argument is fine as far as it goes, but he misses a much more basic point. The total number of compys on the island is 65. If you were to measure the heights of 65 adult people, randomly selected, you would assuredly not find a smooth, perfect bell curve of the type the computer depicts. As with infant mortality, it's a matter of chance and numbers. Given a relatively small group of people, you expect to find an approximate bell curve for their heights, but one with a lot of variation in it. If you measured the heights of 650 people rather than 65, you'd expect something much closer to the mathematically ideal bell curve; measure 6,500 people, and it would be closer still. Likewise, the heights of 65 compys give you a rough, approximate, wiggly version of the bell curve, not the perfect curve Malcolm is shown. The fact that he doesn't pick up on this point hints at a weak spot in his mathematical knowledge (although, as a matter of fact, it's not altogether unusual for ultratheoretical mathematicians to be ill at ease in the messy world of statistics).

There's one more problem here. If you measured the heights of several thousand people, you would expect to get something closely approximating a bell curve, but only if you were careful to exclude children from your survey. (Properly speaking, you would have to do separate surveys for men and women, too, since the sexes have slightly different average heights and if you lumped them in one graph you'd get two slightly different bell curves smeared together.) But the computer is tallying the heights of all the compys, and because this is a new population, a lot of youngsters and juveniles as well as adults are undoubtedly included. In that case, you'd expect to get an excess of short compys, corresponding to the less than fully grown part of the population—representing another distortion of the precise bell curve the computer comes up with.

Looking at the graphs and tables, Malcolm insists that he can prove that dinosaurs have escaped from the park. The lawyer from the Ingen Corporation asks him to explain why he thinks so. "The mathematics are so self-evident they don't need to be calculated" is Malcolm's unhelpful answer. It's a pity he doesn't give us some idea of what his calculations are, because if he can prove mathematically that animals are bound to escape from a nature preserve, it would be a notable discovery. What he seems to be saying is that the designers of Jurassic

Park are attempting to re-create an entire functional ecosystem, not just a zoo of individually penned animals, and that because ecosystems are inherently complex and unpredictable, it's inevitable that things will go wrong. Part of that argument is reasonable, but leaping from there to the idea that dinosaurs are bound to escape, and that the computer records prove they have already escaped, takes some doing.

In one sense, Malcolm is quite right: the designers of Jurassic Park are foolish to imagine that they can re-create an authentic dinosaur ecosystem and expect it to remain stable. Especially in small groups, huge swings in population can be expected. But Hammond and Henry Wu acknowledge this problem: in Jurassic Park, some of the animals have to be kept separate from others, and food for many of them must be brought in from the outside. So there's no question that Jurassic Park is more like a large zoo or a wildlife park than a true re-creation of dinosaurs in their natural habitat. And if you're keeping track of every animal—feeding them, sequestering some in large enclosures, and so on—you have as much control over the animals as your resources and your competence allow. And in fact when things do go wrong in Jurassic Park, it's ultimately because of Dennis Nedry's sabotage.

Malcolm's argument seems to be that chaos theory somehow demands the presence of a fatal flaw, represented here by Nedry's sabotage. But how, exactly? If you wanted to come up with a mathematical model of a dinosaur ecosystem, you would devise equations representing the rate at which each species reproduced, died, ate other species or was eaten by them; and there would also be external factors such as the provision of additional food, the probability of disease, the likelihood of death from exposure to bad weather, and so on. Ecologists can and do put together models of this sort, to get an idea of the likely variation of species populations over the course of time.

But such a model is necessarily limited to the things you put into it. You may find that the animal populations behave in strange and unpredictable ways, and such behavior would indeed be a manifestation of chaos. But there's simply no way for such a model to spontaneously develop an entirely new element, such as the presence of a saboteur. If the only things you've put into your model are dinosaurs, Dennis Nedry can't spontaneously appear out of nowhere. If you did decide to allow for the presence in your model of someone like Nedry, then you're the one predicting that a saboteur might show up, and his appearance has nothing to do with chaos theory.

Malcolm's argument about the vulnerability of the computer program that runs the park does make some sense, however. When President Ronald Reagan proposed his "Star Wars" defense some years ago—an enormous system of missiles, laser beams, and other devices intended to strike down any nuclear missile attack—critics pointed out the impossibility of building and testing the computer programs needed to coordinate the system. You'd have to write huge volumes of software with the capability of tracking thousands of incoming missiles simultaneously and targeting them with countermeasures, but it would be impossible to test the system in advance against all imaginable attacks. The first time the Star Wars software would be fully tested would be the first time it was used—and then you wouldn't get the chance to fix it if it didn't work.

In a broad sense, chaos theory holds that a sufficiently complicated software system can never be thoroughly tested, because the number of ways it can behave becomes unaccountably large. So Malcolm is right to suggest that Jurassic Park's software is likely to behave in unexpected ways—just as a new word-processing program has bugs its designers failed to predict or catch. But it was not software failures that caused disaster to strike Jurassic Park. It's true that because of an elementary programming error, the system fails to detect that the dinosaurs have begun to breed—that their numbers are increasing. But Malcolm catches that error soon enough so that it might have been fixed had Nedry not disabled the entire system, the security fences along with it. Once again, it's sabotage, not chaos theory, that brings catastrophe.

On the other hand, there's an easy way to predict Nedry's role in the story: there has to be a bad guy! That's completely predictable, and, therefore, has nothing to do with chaos theory.

In *The Lost World,* Ian Malcolm has a bigger theme: chaos theory can also explain how dinosaurs became extinct! He indulges as well in a number of asides about the significance of "complexity" as an extra ingredient in evolution—one that must be taken into account along with genes and sex and mutation and selection, all those traditional elements of biology. As usual, it's a bit hard to decipher Malcolm's pronouncements; often, he starts explaining his ideas only when he's been given a large dose of morphine to counter the pain incurred by repeated dinosaur attacks. But we can sort out two main themes.

The first has to do with the difficulty that many people have in understanding how evolution—the random shuffling and mutation of genes—can lead to such superbly specialized biological structures as eyes. You need a complete working eye in order to be able to see anything, so the argument goes, but evolution must necessarily have built up the eye bit by bit. In that case, what were the primitive precursors of our excellent eyes, and what good were they? Because if they were no good for anything, evolution would not have favored them. This argument is a modern variation on the old "argument from design" by which philosophers and theologians wanted to prove the necessity for a divine being. The eye is so complex that it's impossible to understand how evolution could have produced it, therefore someone must have designed it.

Malcolm gives us a version of this argument but hastens to deny that he's any sort of creationist; instead, he hints mysteriously, "other forces are also at work." The other forces, needless to say, come from complexity theory. In particular, he talks about the idea of "self-organization," which is what's supposed to happen when complicated, interacting systems made of many interconnected parts all doing their own thing manage to fall into relatively stable forms of behavior or activity. An oft-cited example of self-organized behavior is what happens when you pile up dry sand on a flat surface. The sand will tend to assume a pyramidal shape whose sides have the same characteristic slope. If the slope is too shallow, the sand will continue to pile up and increase the degree of the slope; but if the slope becomes too steep in one place, sand grains will cascade down the side of the pyramid to reduce the gradient to a stable value. In this way, although the individual grains of sand don't actually conspire to produce a certain slope, the sum total of all their independent actions nevertheless produces a slope of fixed gradient. Malcolm seems to be arguing that some similar element of self-organization is needed to explain how complex biological structures such as the eye can evolve from the interaction of numerous individual genes, which have no way of conspiring in advance to produce an eye.

But as the evolutionary biologist Richard Dawkins has argued at length, notably in *The Blind Watchmaker*, there's simply no reason to think that evolution can't make an eye. Even a primitive, poorly functioning eye—even a single cell that can tell light from dark—would be of use to an organism that previously had had no sense of vision at all. And if a single light-sensing cell is useful, then a group of cells that can distinguish degrees of light and dark, or sense the direction that light is coming from, would be even more useful. It is possible, in other words, to imagine a series of progressively more complicated and efficient light-sensing structures being built up, one tiny improvement at a time. Such "eyes" would have value to the organism possessing them and would therefore be favored by evolution.

What's more, some evolutionary biologists have argued that eyes evolved independently in a number of distinct species, several times over during the development of life on earth. This is plausible without resort to a mysterious "self-organizing" conspiracy among genes. Eyes give the creatures that possess them such a huge evolutionary advantage that the steps toward building an eye have been strongly favored whenever and wherever they happened to crop up.

The other point to bear in mind, as Dawkins emphasizes, is the sheer amount of time that evolution has had to work with: hundreds of millions of years; tens of billions of generations, each a little different from the one that preceded it. We may not be able to imagine a human eye appearing from nothing in, say, a 1,000 generations—but 1,000 human generations is only 20,000 years or so, which barely takes us back to the days of our cave-dwelling ancestors. In evolutionary terms, that's no time at all.

The other point Malcolm keeps harping on in *The Lost World* is more specific. He believes—or hints that he believes—that the reason the dinosaurs died out was because of some sudden and damaging change in their behavior. The whole ecology of dinosaurs forms a complex system, and such systems, Malcolm explains, can behave in abrupt and unpredictable ways. As it happens, there are some genuinely interesting ideas that have emerged from the mathematical study of ecosystems, incorporating some of the insights of chaos theory.

Originally, the theory of chaos or complexity arose from studies of the weather. Because a weather system is so complex—contains so many interconnecting parts, all feeding back on each other in complicated ways—a tiny change in one place can have enormous consequences elsewhere. Hence the by-now-clichéd idea that a butterfly flapping its wings in China can trigger a hurricane in the Caribbean. The reason that complexity leads to unpredictability is that weather patterns differing in only the most minuscule ways can develop, a few days later, into entirely disparate systems—dry and balmy in one case and fierce and wet in another. Because it's impossible to know the initial conditions in perfectly accurate detail, any prediction is likely to go wrong.

A similar line of reasoning comes up in the study of ecosystems, which are interacting populations of animals and plants. For example, think of a grassy region, populated by rabbits that eat the grass and foxes that eat the rabbits. If there's a sudden boom in the rabbit population, the foxes will do better, and their population will rise. But more foxes eat more rabbits, so the number of rabbits will go down again. At the same time, the temporary boom in rabbit population will decrease the amount of grass, because the rabbits will eat more. And as the ground is more and more denuded, the rabbits will starve, and without ground cover it also may be harder for foxes to hide and protect their infants from predators, so that infant mortality will rise among the foxes. And so on. It used to be thought, in a general kind of way, that any system of this sort must settle into some sort of long-term stability, in which the rabbit and fox populations and the grass coverage all stay about the same. And it was usually thought, too, that the more complex the ecosystem, the more *likely* it was to find stability, because of the greater number of interconnections and feedbacks—which were thought of as "checks and balances" ensuring long-term equilibrium.

But that turns out not to be the case at all. Even in simple systems—such as grass, rabbits, and foxes—mathematical modelers have discovered that populations can go up and down irregularly forever, never settling into any stable state. This is an example of chaos: the system never settles down, never becomes predictable.

Even now, it's an unresolved question as to whether an increased degree of complexity makes an ecosystem more chaotic or less chaotic. There may be no general theory that can explain every possibility. This sort of argument, abstract as it may seem, has had some real consequences. The United States Forest Service, for example, has recognized that attempts to put out forest fires as quickly as possible may be an effort to impose more stability on a wilderness ecosystem than is good for it; by putting out all fires, you actually end up increasing the chances of a truly disastrous conflagration in the future.

Malcolm's theory about dinosaur extinction, as far as we can figure it out, runs something like this: Dinosaurs were so dominant, for so long, that in effect they ceased to evolve. They became so well attuned to their environment and to each other that there was no longer any evolutionary pressure for them to change. Malcolm believes that the same is true of the earth's human population today: as a species, we have become so numerous and powerful that we control the environment and are no longer controlled by it. Since we have no natural enemies anymore, we no longer evolve. Survival of the fittest no longer applies to us, because there is nothing to test us.

But this apparent triumph, Malcolm argues, conceals danger. If the dinosaurs became so dominant that they reached a kind of plateau of evolution, then they also became inflexible and unable to respond to change. As a result, even a tiny change could completely destroy them, because they could not adapt to it. This would be the butterfly effect on a grand scale: some insignificant event would trigger a spiral of ever-growing consequences, resulting eventually in the demise of all the dinosaurs. What that triggering event might have been is impossible to say—and immaterial, anyway. Malcolm suggests that a single predator species dying off somewhere allowed a prey population to get out of hand, and that this could have led to global catastrophe.

Oddly, this supposed modern theory of dinosaur extinction is reminiscent of the old Victorian idea that over the course of many millions of years the dinosaurs became slow, sluggish, and complacent, and therefore unable to resist change. It's a bit like the decline and fall of the Roman Empire: The Romans fought their way to the top, vanquishing all their enemies, and once they had established their dominance, they sat around on soft furniture eating grapes and drinking wine and becoming easy targets for the barbarians.

In reality, the idea that the dinosaurs ceased to evolve is hard to sustain. In their almost 200 million years on earth, many dinosaur species became extinct and many new ones arose. Even toward the end of the Cretaceous, 65 million years ago, new dinosaur species continued to appear. Crichton acknowledges that the comparatively late-appearing raptors were smarter and stronger than most of the dinosaurs that came before—which seems to undermine the idea that dinosaur evolution had run into a wall.

Despite everything Malcolm says, it's hard to see how some change of behavior could have been so radical that every single species of dinosaur went from prosperity to extinction.

The extinction was of huge magnitude and scope. Malcolm's suggestion that humans could be driving themselves to extinction through their own behavior may be faintly plausible, but is it really likely that the demise of *Homo sapiens* would bring in its wake the demise of every last mammal on earth? That's the equivalent of what Malcolm is proposing for the extinction of the dinosaurs.

And it's also hard to see how Malcolm hopes to learn anything useful about dinosaur behavior by studying relatively small numbers of animals, unnaturally introduced into a crowded and artificial environment.

Malcolm's theory of dinosaur extinction is sharply challenged right at the beginning of *The Lost World*, when Richard Levine is listening to Malcolm give a lecture on his work. Levine observes that dinosaurs apparently became extinct all around the world at the same time, and asks why, if unpredictable changes in dinosaur behavior led to their demise, those changes should have happened everywhere, in many different places and environments, all at once. Unfortunately, Malcolm just changes the subject. It's apparently a question for which he has no answer.

EPILOGUE

COULD WE?
SHOULD WE?

> *Scientists are focused on whether they can do something. They never stop to ask if they* should *do something.*
>
> —IAN MALCOLM

In 1944, scientists figured out that a molecule called DNA carried hereditary information. In 1953, James Watson and Francis Crick worked out the structure of the DNA molecule and thereby explained the means by which it replicates. Today, scientists have the ability to manipulate little pieces of DNA—genes—and thus affect for good or ill the genetic constitution of whole animals. Much experimentation has been done on plants, bacteria, worms, and fruit flies. Mice have been deliberately created with genetic disorders that mimic similar problems in human beings, in the hope of finding a solution to those human disorders.

Plant breeders have been manipulating crops for centuries by crossing different varieties to create new forms more resistant to disease, more tolerant of drought, more productive. But now scientists can put genes from one plant species into another—the equivalent, if it were possible, of interbreeding tomatoes and bacteria. To scientists, this kind of precise genetic engineering may represent the inevitable refinement of processes that humans have been engaging in ever since they became farmers. But to the general public, the idea of adding and cutting out single genes to control highly specific aspects of a plant or an animal—and, even more, the insertion of genes from one species into another— sometimes seems the first step in transforming a frightening science fiction story into reality.

One day, scientists hope, it will be possible to correct such common and disabling genetic disorders as cystic fibrosis or sickle-cell anemia by deleting or inactivating the faulty gene in a person's genome and replacing it with the correct version. Who could argue against such a cure? In the meantime, we find ourselves frequently in the position of being able to detect such disorders without being able to cure them. Predispositions to various cancers and other life-threatening diseases can now be detected in a fetus. Parents can be screened and told that their children are at high risk for this or that problem. That may leave parents with some difficult choices, but at least the scientific facts are unambiguous. But what happens when all a doctor can tell you is that your child will have an increased risk of some heritable disease but may never develop it? What if a child will be a few inches shorter than average, or has a tendency to obesity? Many of the factors that genes influence may be conditions that society deplores but that are also a normal part of the variable makeup of the human race, not medical disorders in need of correction. The ability to understand and interpret the genetic code has the potential for good and bad (something often true of developments in science). And when it comes to making ethical or moral decisions, scientists are no better or worse than the rest of us. As these new forms of knowledge become public, the public will have to ponder some weighty matters—matters that may make the question of whether or not we ought to re-create dinosaurs seem trivial.

You'll have realized by now that making dinosaurs from ancient scraps of preserved DNA is no easy matter. It may sim-

The "quicken"—a quail-chicken hybrid created for transgenic bird experiments.   (*Drawing by Mick Ellison*)

ply be that no one will ever find any dinosaur DNA at all, or never any more than a few scraps of it. And even if sufficient dinosaur DNA is found, there are countless difficulties and complications in creating a dinosaur—enough to make the task essentially impossible. But the sorts of problems that remain perplexing—how to identify and "mend" unknown genes, how to persuade an egg cell to take up a foreign genome, how to manipulate the genetic characteristics of a growing embryo—are those that scientists are working on right now, in many different contexts, in laboratories around the world.

The work that's been done up to now in creating genetically engineered organisms and releasing them into the environment is modest, compared to making a dinosaur, but nevertheless has been a battleground both legally and politically. Some years ago, researchers in California inserted a cold-resistance gene into *Pseudomonas* bacteria. Eventually, these engineered bacteria were sprayed onto strawberry plants in open fields, in order to ward off frost damage. The enterprise had to overcome enormous opposition—particularly because this was the first deliberate attempt to release altered genetic material into the environment.

In matters of genetic engineering, it's public opinion that seems to win the battle in the end. If frost-fighting bacteria make strawberries abundant year-round, it's likely that concern about possible technological ill effects will fade.

If ever it does seem possible to revive dinosaurs, public enthusiasm may make it impossible for cooler regulatory heads to hold the project up, or even debate its ethical implications. Undoubtedly, there will be some people who deplore the notion of a Jurassic Park—who find it offensive, for one reason or another, for scientists to rebuild what nature long ago rendered extinct. Others will decide the project is too dangerous to attempt. But John Hammond was surely correct in thinking that huge numbers of people will find the idea, whatever perils it may bring, so fascinating, so awe-inspiring, that they will provide enthusiastic and essential support.

Fifty years from now, it may be that many of the technical obstacles to dinosaur re-creation—obstacles that seem insurmountable today—will have been overcome through the efforts of scientists pursuing all kinds of other goals. Yet if we ever reach the point where we have the technical ability to create dinosaurs, we may find that the ethical problems involved in resurrecting an extinct creature will pale in comparison to the quandaries we will encounter because of our ability to manipulate the tiniest genetic elements of living creatures—most particularly, ourselves.

INDEX

Page numbers in *italics* refer to illustrations.